Leadership Astronomy

Leadership Astronomy

Seeking Not Old-fashioned but
God-fashioned Leadership

Bertrand Millhouse

Best wishes, Judith

Bertrand Millhouse

VANTAGE PRESS
New York

All biblical quotations were taken from the King James Version of the Holy Bible, unless otherwise noted.

FIRST EDITION

Published by Vantage Press, Inc.
516 West 34th Street, New York, New York 10001

Manufactured in the United States of America
ISBN: 0-533-11848-4

Library of Congress Catalog Card No.: 96-90008

0 9 8 7 6 5 4 3 2 1

Dedication

To my beloved mother, Mrs. Beatrice Millhouse, who taught me how to pray when I was just a little child, and to the memory of my beloved deceased father, Mr. Bertrand Weil Millhouse, who taught me responsibility and the significance of commitment during my early adolescent years. My parents were devout Christians who evangelized our home. They raised fourteen children and they encouraged us all to attend Sunday School every Sunday. We really enjoyed going to the Lord's house—being in God's presence. Above all, I thank the God of Abraham, Isaac and Jacob for giving me parents who were model leaders in the home. They taught us Christianity—righteousness. They taught us the biblical golden rule, "Do unto others as you would have them do unto you." Knowing that "a house divided against itself cannot stand," they taught us to love one another as they loved us. With very little they did very much because they put their little into the hands of a gracious God.

Contents

Preface

Many years ago King David said, "I will look to the hills from whence cometh my help. Thy word is a lamp unto my feet and a light unto my path." Indubitably, the king was a spiritual astronomer because he kept his eyes focused on the hills of Zion. Through his telescopic eyes of faith, he looked above the material universe and beyond the horizons to seek counsel and guidance from the Lord. Whenever a leader lifts his eyes to the hills for help, he engages in spiritual astronomy. Ps. 121:1 and 119:105 are the keys that unlock the doors to spiritual leadership astronomy. However, when a leader decides to look no longer to the hills for help and relies exclusively on his own sufficiency, he falls from grace. Moreover, only the ancient of days can restore his joy of salvation and godly servantship. The Holy Bible says: "Grow in grace and in the knowledge of our Lord and Savior Jesus Christ."

Now let's take a quick look at the astronomy of the prophet Daniel in Dan. 12:3: "They that be wise shall shine as the brightness of the firmament and they that turn many to righteousness as the stars for ever and ever." Daniel, the beloved prophet, was unquestionably a spiritual astronomer. Though taken into Babylonian captivity, he prayed three times per day for Israel and he frequently fasted as well. Daniel developed an intimate relationship with God and God gave him favor with the king. Hallelujah! Hallelujah!

Finally, let me introduce you to the spiritual astronomy described by our Lord Jesus. In Rev. 22:16, Jesus says, "I am the root and offspring of David and the bright and morning star." In other words, Jesus is our astronomy. We study him through our built-in telescope of faith. Our faith cometh by hearing and hearing by the Word of God. The Lord says He's going to come back for a church that is without spot or wrinkle.

In order to fulfill His purpose, the Lord needs spirit-filled leaders who will look only to the hills of Zion for counsel and guidance. God needs bold leaders who will not compromise the

Word of God to appease those who have itching ears. He needs leaders who will say to the people, “I came to tell you what thus saith the Lord.” God desires model leaders who will lead the sheep by example. If the leader keeps his eyes focused on Jesus, he will be a model leader indeed! Leaders must learn to say, like the apostle Paul, “Follow me as I follow Christ.” Jesus never fails, and God is able to keep that which we commit to Him in Christ Jesus. So the bottom line is watch Jesus, the star, and you'll be a victorious leader pleasing unto God. My friends, “Let this mind be in you which is in Christ Jesus.”

The Lord of Hosts is able to make your enemies be at peace with you. He even has the hearts of the kings in His hands. The judgment of God begins in the house of the Lord. As we look around us, we see that there is a shaking going on in the house of God. Therefore, I pray, “Lord of heaven and earth, thou who standeth in the congregation of the mighty, use this book to turn the hearts of arrogant, prejudiced, egotistical, and disobedient leaders to Christ. Then and only then will they please thee. I pray also that you will use this book to encourage and edify spirit-filled Christian leaders throughout the world. Thou art the ruler of all nations! Let thus-saith-the-Lord leaders rise up in holy boldness and let them prevail, Almighty God! You said in your Word that in righteousness we can be as bold as a lion. Lord, by faith I thank you for hearing my plea. In Jesus' holy name, I pray. Amen.”

Acknowledgments

I wish to express my profound gratitude to our triune God for giving me through His Holy spirit the vision and inspiration to write this book. I thank God for the vision and for anointing my fingers to scribble these lines. To the Father, Son, and Holy Ghost: Thank you, thank you, thank you. Special thanks to Rebecca Falero for her typing service and to Joey Rodriguez for his help in the artistic design.

Leadership Astronomy

Keys to Understanding Leadership Astronomy

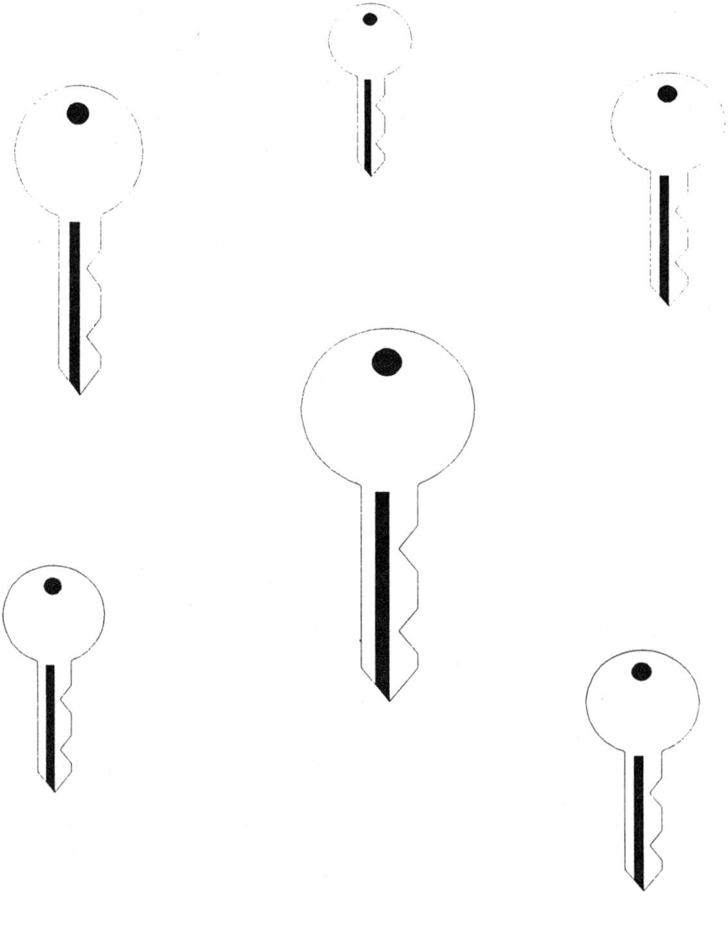

I

Keys to Understanding Leadership Astronomy

In the Book of Proverbs, the Lord says, "Incline thine ear unto wisdom, and apply thine heart to understanding; happy is the man that findeth wisdom, and the man that getteth understanding. For the merchandise of it is better than the merchandise of silver, and the gain thereof than fine gold" (Prov. 3:13–14). That means in all our "getting" we should get an understanding, for understanding is more precious than silver and gold. If a man gained all the silver and gold of this world but failed to secure understanding, he would surely lose it. Lack of understanding is a tragedy! Therefore, I am going to devote chapter 1 of this book to understanding the astronomy of Christian leadership. I am going to give you the keys that unlock the doors of the subsequent chapters that you might taste and see that the Lord is good.

Who institutionalized leadership on the earth? And what are the keys to leadership astronomy? The most high God is unquestionably the instituter of leadership on the earth. Let's take a look at Gen. 1:27–28: "So God created man in His own image, in the image of God created He him; male and female created He them. And God blessed them and God said unto them, be fruitful, and multiply, and replenish the earth, and subdue it; and have dominion over the fish of the sea, and over the fowl of the air, and over every living thing that moveth upon the earth." Within these Scriptures we can clearly see the origin of leadership on the earth. During the early stages of human history, we see that divine and human leadership coexisted. Daily the Lord came down from His holy hills to fellowship with His human creation in the Garden of Eden. He gave Adam and Eve instructions and directions. Above all, that garden was filled with the love and glory of the Lord. It was a paradise! Unfortunately, the human leadership of God failed Him and got us all into a world of trouble. Be that as it may, earthly leadership has its roots in the Garden of Eden of the Old Testament. There, God also reveals to us something about His divine leadership. Many of you have read the Holy Bible in its entirety from Genesis to Revelation. Perhaps some of you have not. Therefore, I say when you will have read the story of Eden and the account of God's unfailing love and mercy in Genesis (Gen. 3:15), you will conclude that divine leadership never fails.

Earthly leadership must hinge on the divine. The Lord says He will be a buckler to them that walk uprightly. Inaudibly say, "The Lord will be a buckler to them that walk uprightly." Now if you're reading this book in privacy, say it *audibly*. That's right, say it again. I want you to meditate on the Word. That way, you'll never forget it. It will become a part of you. Ps. 1:2 says: "But his delight is in the law of the Lord; and in his law doth he meditate day and night." We need to meditate on the Word of God day and night in this old sin-sick world. We need to praise God in season and out of season. In doing so we indulge in spiritual astronomy. Furthermore, when we meditate on God's Word we medicate our spirits. Meditation = Medication. Now that's the divine formula the Holy Spirit told me to give to you: meditation = medication, medication for the soul.

The following fifteen terms are the keys to understanding leadership astronomy. Though most of them are secular words, I'm going to use them as vehicles to drive home spiritual points. If you meditate on these terms, I guarantee you will acquire insight into Christian leadership astronomy.

Now let's begin our word analysis:

Spiritual astronomy: It is a systematic study under the inspiration of the Holy Spirit of the Word of God applying hermeneutic techniques or methods. It's the art of taking a telescopic look at the Lord Jesus, the bright and morning star, thereby observing His attributes and applying His teachings.

Spiritual astronomer: A spirit-filled believer who looks to God for divine guidance or leadership. He keeps his eyes focused on Jesus, who is the way, the truth, and the life. The Word of God says, thou shalt keep him in perfect peace whose mind is stayed on thee and if thine eye be single thy whole body shall be full of light. The spiritual astronomer says, like King David in Ps. 112:1; "I will lift mine eyes to the hills from whence cometh my help." Lifting your eyes to the hills of Zion for help makes you a spiritual astronomer. The Word of God says God is a spirit and they that worship Him must worship Him in spirit and in truth. You can only see God through your telescopic lenses of faith.

Telescope: The Lord says in the Word, "Heaven is my throne and the earth is my footstool." Knowing heaven is so far away, God created man with built-in telescopic lenses of faith, which enable him to draw nigh to God. The Bible says the fool says in his heart there is no God yet the Heavens declare His glory. Faith cometh by hearing and hearing by the Word of God (Rom. 10:17). The physical telescope is an optical instrument for making distant objects appear nearer and larger, but the spiritual telescope is our optical instrument of faith, which enables us to know God through faith and to see that the Kingdom of God is at hand.

Microscope: It's a spiritual inner optical instrument used by the spiritual astronomer or leader when he's inspecting objects too small to be seen distinctly and in detail with the naked inner eyes, figuratively speaking. The issues or situations to be studied are not distant. They're right at hand and they only need to be magnified.

Macrowisdom: It's the application of knowledge in a situation that is large enough to be studied and identified with the naked eyes. One does not have to take a microscopic look at the subject. Instead the astronomer uses macrowisdom and common sense. Don't forget "common sense" is not a common thing!

Microwisdom: It's the application of knowledge in situations that are too small to be seen clearly and understood with the naked eyes. The soul cries to the hills for wisdom and understanding. The Lord said, if any man lacks wisdom, let him ask the Lord who giveth wisdom liberally and upbraideth not.

Astrophotography: In a secular sense, astrophotography is the photography of the stars and other celestial objects, but in a spiritual sense, it is what we see through our eyes of faith. "Eye hath not seen, ear hath not heard, neither hath entered into the heart of man what God has in store for those who love him" (1 Cor. 2:9).

Christian leadership: It's the art of directing men in the paths of righteousness. It's teaching and showing them by your lifestyle model example that Jesus is the Way, the Truth, and the Life and that no man comes to the Father but by Him. It's saying like the Apostle Paul, "Follow me as I follow Christ."

Leader: An anointed servant of God who directs or brings spiritual enlightenment to the sheep of God and to lost souls in this sin-sick world. He is a commander. The Hebrew word is translated as prince (I Sam. 25:30). He feeds the sheep of God through the empowerment of the Holy Spirit. The Bible says, "Beautiful are the feet of men who preach the Gospel of Peace" and "When the righteous ruleth the people rejoice."

Abraham: The father of spiritual astronomy, the father of faith (Rom. 4:16–25). Now "faith is substance of things hoped for, the evidence of things not seen" (Heb. 11:1). God calleth those things which be not as though they were.

The Lord Jesus: Jesus is our spiritual astronomy because He says (Rev. 22:16) "I am the root and offspring of David and the bright and morning star." And in John 14:6, Jesus says, "I am the way, the truth and the Life. No man cometh to the Father but by me." Jesus is our astronomy; He is our guiding light. The Lord paid an astronomical price to redeem fallen man. He gave His life.

The heart: It is the center of your thoughts. It's your mind. It's you. Prov. 23:7 says, "For as he thinketh in his heart so is he. Eat and drink says he to thee but his heart is not with thee." Out of the heart are the issues of life (Prov. 4:23). The Bible says, "The Word of God is quick and powerful and sharper than any two-edge sword piercing asunder to the bones and marrow and is a discerner of the thoughts and intents of the heart" (Heb. 4:12). With these thoughts in mind, it is crystal clear that the condition of your spiritual heart plays a key role in your leadership. Please remember that before you can do spiritual astronomy and leadership, it is necessary to let God do your heart surgery. Jesus said, "Be ye transformed by the renewing of your mind."

The telephone: The Word of God says we are fearfully and wonderfully made. That means we are all unique. No man is the same as another. God doesn't make carbon copies of people. The Lord has designed within our souls a telephonic communication system, which enables us to pray to Him without ceasing and to praise Him in season and out of season. The telephone calls to our Lord don't cost one dime. They're free because Jesus paid it all on Calvary when He gave His life as an atonement for our sins. He purchased our liberty.

The cablegram: In Mark 16:15, Jesus said, "Go ye into all the world and preach the Gospel to every creature." The Lord Jesus sent the cablegram to His church many centuries ago. Nevertheless, we must keep responding to it because souls of Macedonia and other nations are still crying out for help. Souls across the sea are crying, "Come and teach us the ways of the most High God." I see a cablegram in Mark 16:15.

Hermeneutics: It is the art or science of interpreting the

Sacred Scriptures. The hermeneutic methods of Scripture analysis are also applied in spiritual astronomy. Please note that spiritual astronomy and hermeneutics are inseparable.

Chapter I has provided you with the rudiments of leadership astronomy. Into your hands also are the keys that unlock the doors to the following chapters. Let us go forth.

Divine Steps to Ministry Preparation

To everything there is a season,
and a time to every purpose
under the heavens. Ecc. 3:1

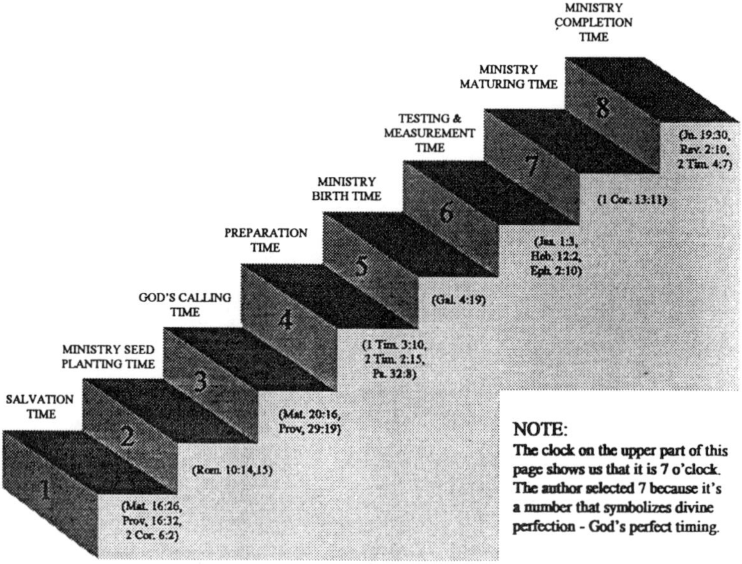

NOTE:
The clock on the upper part of this page shows us that it is 7 o'clock. The author selected 7 because it's a number that symbolizes divine perfection - God's perfect timing.

II

Divine Steps to Ministry Preparation

All journeys (whether short or distant) begin with a single step. If you decided to walk a marathon of twenty-seven miles, it would still begin with a single step. That a journey begins with a single step is an indisputable universal truth. However, completion of the ordeal is not determined by how big or small steps you take but by the quality and consistency of the steps. We must not forget that the steps and measurement of a righteous man are ordained by a Holy God. Bear in mind that the race is not to the swift, nor the battle to the strong. Don't turn back because the crown awaits the conquest. If you ask Him, God will give you "Amazing Grace," which will enable you to endure as a good soldier, and He will give you the crown at your journey's end.

In Eccl. 3:1, the Lord says, "To everything, there is a season, and a time to every purpose under the heaven." This Scripture demonstrates that our God is a God of order. He orchestrates the seasons and time. Look at the earth and look at the heaven. Both declare His glory. There is order in God's universe. There is order indeed! Now in this chapter, we're going to talk about the divine steps in ministry preparation. Let us begin our discussion. Note, I said "discussion" because I expect you to read, meditate on, and discuss this book with inquiring minds throughout the world.

Though I am going to explain shortly God's divine plan for ministry preparation, I must warn you that God also has a permissive plan. Take a good look at Rom. 11:29 and you'll discover it: "For the gifts and calling of God are without repentance." And how about that dialogue between the Lord Jesus and Peter after Peter's calling but before his conversion: And the Lord said "Simon, Simon, behold, Satan hath desired to have you, that he may sift you as wheat; but I have prayed for thee, that thy faith fail not; and when thou are converted, strengthen thy brethren. And he said unto him, Lord I am ready to go with thee, both into prison, and to death. And He said, I tell thee Peter, the cock shall crow this day, before that thou shalt thrice deny that thou knowest me" (Luke 22:31–34).

Now let's take a walk up the divine steps to ministry preparation.

As you can see from the diagram there is a *time* for stepping, and there are eight basic steps in ministry preparation. We're going to do a spiritual analysis of each step in each time zone as we walk up the mystic path:

Salvation Time

A leader's salvation is the cornerstone to his ministry. It is something that is essential and indispensable. Moreover it is a stone representing the nominal starting place in the construction of a monumental ministry. As we look around us, we see that there are two types of leaders in this world: (1) the godly leaders and (2) the so-called leaders. The godly leaders are those who honor Jesus as Lord. They see Him as the bright and morning star, and He is unquestionably their spiritual astronomy course. In Him, they live and move and have their being (Acts 17:28). They are fearless leaders who will not compromise the Word of God to appease anyone. Their ship is anchored in Christ Jesus, and they will never forget their first love. Godly leaders have the joy of salvation and they rejoice to see the salvation of others.

On the other hand, the so-called leaders are those "having a form of Godliness but denying the power thereof" (2 Tim. 3:5). To them, winning the applause of men is more significant than winning the smile of God. They compromise the Word of God to please those who have itching ears. They have forsaken their first love who will judge all men in eternity. So-called leaders are men pleasers. Neither the joy of salvation nor the Holy Spirit's ventilation can be observed in their barns. The theme of the so-called leader is "what's in it for me." He is motivated only by personal gain and prosperity.

Salvation is the first step in a leader's monumental ministry preparation because without it, he would have no joy. The "joy of the Lord" is the leader's strength. Salvation first is God's divine order. In 2 Cor. 6:2, the Lord says, "Behold now is the acceptable time; behold now is the day of salvation." Notice how the Lord uses the adverb "now" in the passage. He uses the word twice and his repetition denotes urgency and prioritization. Now is a time period. It means at this moment or the present time. The leader should understand that it would not profit him anything if he gained the whole world and lost his own soul (Matt. 16:26). He should also understand that he that ruleth his spirit is better than he that conquereth a city (Prov. 16:32).

Ministry Seed Planting Time

After God's prospective leader receives Christ as his Saviour and Lord, the Holy Spirit moves miraculously within his spirit to turn a heart of stone into a heart of flesh. In the ministry seed planting time zone, God does major heart surgery; in fact He does a heart transplant. It is evident that the prospective leader needs a new heart. Before the Lord plants a seed, He cultivates the soil and then irrigates the land. After the soil preparation, God plants the seed for leadership harvest time. "And how shall they hear without a preacher? And how shall they preach, except they be sent? How beautiful are the feet of them that preach the Gospel of Peace, and bring glad tidings of good things" (Rom. 10:14–15). Here we can easily conclude that our God is also an agriculturist. He knows all about seed planting. Trust Him and His infinite wisdom because only He can plant a seed and cause it to grow. Unless the Lord plants the seed, it will not grow. What is the seed? It is one's innate desire to minister unto the Lord and to minister unto men in whatever capacity inspired by the Holy Spirit of God. It's a seed of ministry.

God's Calling Time

When the seed has sprung up and is ready for harvest, our sovereign God sounds the alarm. He calls. According to Matt. 20:16: "Many are called but few are chosen." Perhaps you are wondering why does the Lord call many but choose few. Do I have the answer for that! The answer is simple. Our righteous God chooses few because only a few of those whom He calls respond favorably to His call. Those who respond unfavorably to God are entangled in this system of worldliness and are too busy to say yes to God though they understand the call. In Prov. 29:19 the Lord says, "A servant will not be corrected by words: For though he understands he will not answer. Never say no to God's call. His call is not to your ability but to your availability. Say like Isaiah, "Here am I Lord; send me," and make your calling an election sure.

Preparation Time

In the preparation time zone, the prospective leader receives training and instructions from the Lord. God says to him, "I will instruct thee and teach thee in the way which thou shalt go: I will guide thee with mine eye" (Ps. 32:8). The trainee intensively studies the Word of God so that he might be approved by God and so that he might properly analyze the Word of Truth. The Bible says, "Let them first be proven" (tried or investigated).

Ministry Birth Time

After the prenatal care preparatory period comes the birth of the baby—the birth of the ministry. God says travail in birth again until Christ be formed in you (Gal. 4:19). It is obvious that you will experience some birth pains, but don't grow weary. In Christ Jesus we are more than conquerors.

Test and Measurement Time

In this time zone, the Lord tests you so that this might develop patience within you. "Knowing this that the trying of your faith worketh patience" (James 1:3). "Let us run with patience the race that is set before us, looking unto Jesus the author and finisher of our faith." Though the trying of our faith be with fire, to God our faith in Him is more precious than silver and gold. "For we are His workmanship created in Christ Jesus unto good works, which God hath before ordained that we should walk in them" (Eph. 2:10). Let me hasten to warn you that all through your life's journey you will have tests. The tests are unending, and it is the Holy Spirit of God who guides and enables us to score passing marks on the exams. Say "I can do all things through Christ which strengthens me" (Phil. 4:13).

Ministry Maturity Time

In 1 Cor. 13:11, it is written: "When I was a child I spake as a child, I understood as a child, I thought as a child; but when I became a man, I put away childish things." Oh, leader, use your God-given wisdom. Put away childish things. Avoid foolishness and horseplay. Clean humor inspired by God is acceptable but not foolishness, disrespect, and outright disobedience. Be a man in your thinking, but be like a child in respect to your obedience to God. Grow in grace and in the knowledge of our Lord and Savior Jesus Christ. Add to your faith virtue; and to virtue knowledge; and to knowledge fraternal love. Always remember that "faith without works is dead."

Ministry Completion Time

A leader's mission is completed when he has done all that God has called him to do in this world. Therefore, a leader's ministry does not end when he retires from an earthly position, but his ministry ends at his death. You don't retire until you expire. God doesn't have retirees. Jesus said, "Lo I am with you always, even to the end of the world. Be faithful until death and I will give you a crown of life." It is not over until life is over. Let's take a quick look at the Lord Jesus hanging on the cross at Calvary. His last words were, "It is finished," and He bowed His head and gave up the ghost (John 19:30). The apostle Paul said at his ministry completion, "I have finished my course, now I am ready to be offered." None of us know the day nor the hour that the bell will toll for us. But there is one certainty: if the Lord Jesus should delay His coming, it will surely toll someday. Godly servants will be prepared because they walk and talk daily with God. But the Lord's coming will be a big surprise for the unrighteous. He promised the nonbelievers that He would return to earth in a moment that they least expect. Jesus says in the Word, "Behold I come quickly. Let the righteous be righteous still; and let the unrighteous be unrighteous still."

You have just completed studying your walk up the divine steps in ministry preparation. As you walk up these steps in actuality, let it be known that you're a soldier in the army of the Lord! Leaders in ages past have victoriously walked up these steps. Moses and Elijah walked up these steps; the beloved prophets Daniel and John walked up these steps; the apostle Paul walked up these steps; though He came that the world through Him might be saved, the Lord Jesus walked up these steps for the sake of the human race and was triumphant! Hallelujah! Praises and thanksgiving be to the Lord of Lords and the King of Kings forevermore.

Sound Structure of Christian Leadership

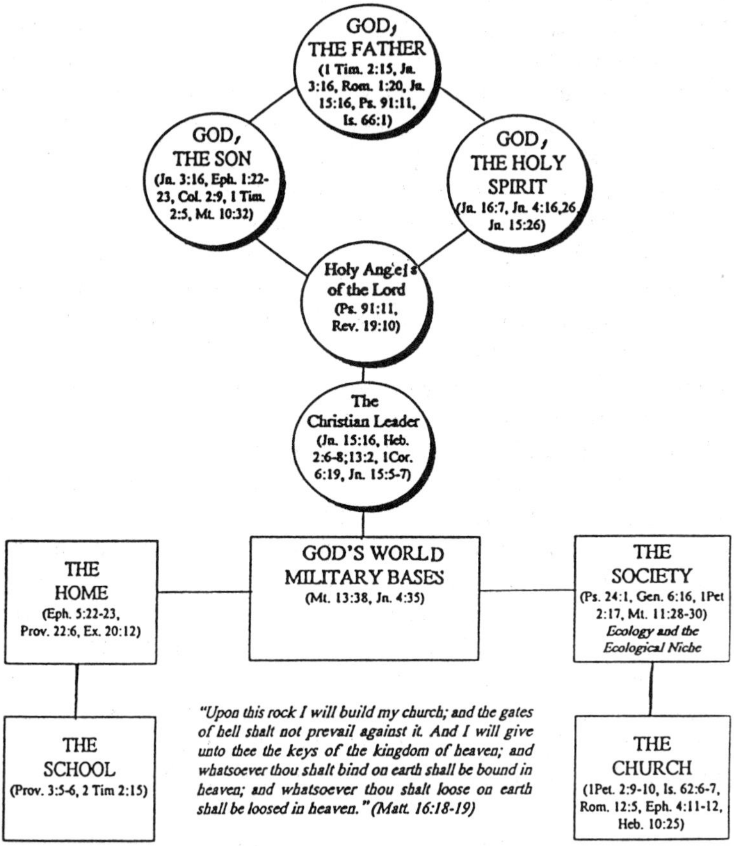

III

Sound Structure of Christian Leadership

On page 15 is a map of the sound structure of Christian leadership. In order to understand, you must study with a *spiritual* mind because the *carnal* mind does not *understand* spiritual things. The map indicates some spiritual truths based on the sacred Scriptures. It tells us that our Triune God plays the major role in Christian leadership. The holy angels play *supporting roles*; and men play *minor roles.* Angels and men are merely vessels through whom God works miracles. This should not surprise you because we serve a mighty God! Our God is omnipotent! He's omniscient; and He's omnipresent. Therefore, we can obviously feel the impact of the invisible, sovereign hand of God in the affairs of this world. Ps. 95:7 says: "For He is our God; and we are the people of His pasture, and the sheep of His hand."

Furthermore, if every pastor ceased his labor, God would still be God; if all Christian leaders and evangelists ceased their labor, God would still be God. If church ushers and deacons ceased their labor, God would still be God. In other words, God is exclusive. He doesn't *need* anybody else. The Lord says in the Word, "I am alpha and omega, the beginning and the end." God has always been in existence and He will exist forevermore. He wasn't created but "all things were made by Him: without Him was not anything made that was made" (John 1:3). In Isa. 45:22, the Lord declares, "I am God, and there is none else." As I stated previously, God is exclusive! However, He *included* men and angels in His universal program. God is *exclusive* but leaders are *inclusive.*

Let us take an in-depth look at the structure and functions of Christian leadership. We shall do an analysis of the specific functions of the Triune Godhead, the holy angels, and the Christian leaders. Some people do not believe in the existence of the Godhead, but they are sincerely wrong. The Holy Bible is our guide and it confirms that there is a Godhead. If God says it, that settles it. In I John 5:7, the Lord says, "For there are three that bear record in heaven, *the Father, the Word. and the Holy Ghost: and these three are one.*" (Italics added.) In Gen. 1:26, God says, "Let us make man in our image, after our likeness." Whose report do you believe, reader? Do you adhere to what the Word of God says, or do you believe *man's unsupported by the Word of God opinion?* I urge you to believe what thus saith the Lord, because "Every word of God is pure, He is a shield unto them that put their trust in Him" (Prov. 30:5). Just as the Body of Christ is one in the spirit consisting of many believers, so is the Godhead one in spirit consisting of three persons with different functions. During His earthly ministry, the Lord Jesus prayed, "Father, make them one as you and I are." Now let's take a look at the major role of the Godhead in Christian leadership.

Leadership Functions of God the Father

In Isa. 66:1 it is written: "Thus saith the Lord, the heaven is my throne and the earth is my footstool. Where is the house that ye build unto me? And where is the place of my rest?" In this verse the word "throne" refers to the seat and symbol of the sovereign authority of the almighty God. "Footstool" is a comfortable design created for the rest of one's feet. God created the earth to symbolize His footstool, and the specific place of His rest is in the hearts of Christian leaders and their followers. God is supreme. He makes the plans and man fulfills them. There is order in God's universe. First Tim. 2:5 says: "There is one God, and one mediator between God and man, the man Christ Jesus." According to 1 Kings 8:30, God hears the supplications of His servants and of His people Israel when they call. He hears them and forgives their sins. The leader is just a lump of clay in the potter's hand. Therefore, he must comply to the wishes of the potter.

It takes tremendous discipline, but God equips all of His chosen leaders. Though Adam failed God, He has given fallen man another chance at leadership because of the redeeming love of His only begotten Son. Had not the Son of God endured the awful suffering of that cruel cross at Calvary, there would be no Christian leadership. The word of God verifies that Jesus paid it all. In James 1:17 it is written, "Every good gift and every perfect gift is from above and cometh down from the Father of Lights with whom is no variableness, neither shadow of turning." The Scripture says, "No man comes to the Son except sent by the Father and no man comes to the Father except sent by His Son." The Father also gives His holy angels charge over His leaders to protect them (Luke 4:10).

Leadership Function of God, the Son

God, the Son, is the mediator between God, the Father, and men. Let me hasten to tell you again that the Word of God says in 1 Tim. 2:5: "For there is one God and one Mediator between God and men, the man Christ Jesus." God is the Son in redemption, and He purchased man with the blood that He shed on Calvary Hill. On the day that Adam and Eve sinned in the Garden of Eden, man fell from grace and became captives of Satan, the serpent. In Gen. 3:15 the Father speaks to Satan, the captor, and tells him, "And I will put enmity between thee and the woman, and between thy seed and her seed; it shall bruise thy head." Through Jesus' atoning death on the cross, He bruised the serpent's head and paid the ransom price for all sinners who desire to receive Him as Lord and Savior. Leaders need to understand that Jesus is the head of the church. Eph. 1:22–23 supports this truth: "And hath put things under His feet, and gave Him to be the head over all things to the church, which is His body, the fullness of Him that filleth all in all." According to the third chapter of Col., the Son sitteth on the right hand of God, the Father. He serves as a divine attorney for all believers. For this reason, the Apostle Paul said in Phil. 4:13, "I can do all things through Christ which strengtheneth me." Furthermore, the Lord Jesus shows the assignments for His leaders to the Holy Spirit of God and the Holy Spirit shows them to the leaders. With reference to the Holy Spirit, Jesus says in the Word, "He shall glorify me, for He shall receive of mine, and shall shew it unto you. All things that the Father hath are mine; therefore said I, that He shall take of mine, and shall shew it unto you" (John 16:14–15).

Leadership Functions of God, the Holy Spirit

Based on John 14:26, God's Holy Spirit teaches, guides, and comforts the leaders. He shows them the will of God for their leadership. He is the illuminator of the mind. Therefore, the leaders who desire to follow Christ will not walk in darkness. He speaks through dreams, visions, and through the Leader's still-small voice within his soul. When we're reading the Word of God, He speaks to our minds and causes us to clearly understand what we're reading. Sometimes the Holy Spirit or Holy Ghost also speaks to us through other servants. He moves in their spirits and gives them a word of knowledge or prophecy, which is for the edification of the believer. Sometimes the Holy Spirit speaks to the leaders through situations experienced. He gives them warnings or advice. Based on Rom. 8:26, the Holy Spirit makes intercessions for us, because He knows all about our infirmities. He intercedes for the saints according to the will of God. Without the anointing of the Holy Ghost, a leader cannot minister with power from on High. The Spirit baptizes and empowers the leader and his followers. According to John 16:8, the Spirit convicts us of our sins; and according to John 16:14, He shows us the divine will of God. Hallelujah!

The Supporting Leadership Roles of the Holy Angels

The Holy Angels of the Lord are heavenly messengers and they are ministering spirits. Our omniscient God has strategically placed them in leadership supporting roles. Ps. 91:11 tells us that He gives His angels charge over saints. Let's examine the exciting roles the angels play:

1. They are God's secret agents and men cannot see them with the natural eyes unless the Lord allows it. They do warfare in the spiritual arena on behalf of the heirs of salvation.
2. They protect us from Satan and his demonic spirits. Eph. 6:12 reveals to us that we wrestle not against flesh and blood, but against principalities, against powers, against the rulers of the darkness of this world, against spiritual wickedness in high places.
3. According to Rev. 19:10, the Holy Angels of the Lord are fellow servants of the Christian leaders. When God chooses, men entertain angels without knowing it (Heb. 13:2). Therefore it behooves us all to treat everybody fair and justly.

For our Lord Jesus says, "In as much as you have done it to the least of one of these you've done it unto me." They minister to God's servants. Most of their labor is done in the spiritual realm, but men receive the blessings of it and sing his praises unto God. In John 4:11, the Word tells us that angels came and ministered to the Lord Jesus after the devil tried tempting Him on that mountain. Just as the Lord loves the saints of God, so do the holy angels. The Bible says there is joy in the presence of the angels of the Lord when a sinner repents. Let us thank and praise the Lord for the wonderful work He does for men through His angels. To God be glory and honor.

The Minor Leadership Role of the Christian Leader

I call the leadership role of the Christian leader a minor one because the work that he does is only a reflection of God's craftsmanship within him. The Lord says, "Let your light shine before men that they may see your good work and glorify your Father who art in heaven." No godly leader is ever born, but he is created by God. God is the architect of the leader's physical body and his spiritual tabernacle of praise. Man's wisdom comes from God; his understanding comes from God; his knowledge comes from God; his skills and talents come from God; man owes all that he is and ever will be to Calvary! As I said previously, man is just a lump of clay in the potter's hands. When man gains an understanding of this spiritual truth, he will surely decrease and the Lord Jesus will increase in his character and in his ministry. In John 15:16, Jesus says, "Ye have not chosen me, but I have chosen you, and ordained you, that ye should go and bring forth fruit, and that your fruit should remain."

Moreover, the leader needs to understand that Jesus is the true vine and the leader is only the branch. Without God's help he can do nothing. So with that thought in mind, the main job for the Christian leader is to lift the banner of Jesus and march! By march I mean he must lead the sheep in the path of righteousness, teaching them that Jesus is the way, the truth, and the life. He should never compromise the Word of God to appease anyone. God wants the Christian leader to be a man of prayer, meditation, and fasting. As a spiritual astronomer, the leader must look to the hills of Zion for help. The leader gets his inspiration for leadership assignments from the Holy Spirit of the Lord. He implements them and establishes them on Christ, the Solid Rock. "What? know ye not that your body, is the temple of the Holy Ghost which is in you, which ye have of God, and ye are not your own. For ye are bought with a price; therefore glorify God in your body, and in your spirit, which are God's" (1 Cor. 6:19–20).

God's World Military Bases

"No man that warreth entangleth himself with the affairs of this life, that he may please him who hath chosen him to be a soldier" (2 Tim. 2:4). The Lord of Hosts is raising up a powerful army of fearless men, women, and youths in these last days. The people of the world cannot see them with natural eyes. Only the people of God are cognizant of this fact. Nevertheless, the soldiers of the cross are marching up to Zion "with the banner of Jesus going on before." The world cannot see God's invisible army, yet it sees the works of the army of the adversary of God, and man and glorifies its evil. The television industry, the movie industry, and the news media are largely responsible for promoting violence, immorality, and anti-Semitism in our decaying society. It will all return to haunt them forever. "Vengeance is mine," saith the Lord.

In John 4:34 the Lord tells us that the fields are white and ready for harvest. The world and they that dwell therein are the fields. Let's take a look at God's key military bases on the battlefield: (1) the home, (2) the school, (3) the society at large, and (4) the church.

The Home

Leadership does not begin in the House of the Lord. For it begins at home. God says in His Word, "For if a man know not how to rule his own house, how shall he take care of the church of God" (1 Tim. 3:5). God has strategically placed the parents in the home as the spiritual leaders and bread winners. The man is head of the house and the wife is his helpmate. God ordained it and that settles it. Disobedience to the will of God is an abomination. It is an attempt by Satan to destroy the family plan, giving a deadly blow to God's divine will. That's why God tells us in the Word that the sin of disobedience is as the sin of witchcraft. Divine leadership training must begin in the home in the family structure. In Prov. 22:6 God says, "Train up a child in the way he should go and when he is old he will not depart from it."

Regardless of what position one holds in society, he grew up in a home. What he is, what he thinks, what he does are all reflections of what he was taught at home and in his environment. Politicians, policemen, government agents, senators, presidents and prime ministers, schoolteachers, mayors of cities and towns all grew up in somebody's house. Their basic characters were formed there. Position or promotion in society does not change one's behavioral pattern. For example, if one was prejudiced before he became a judge, he will be prejudiced afterwards regardless of the oath he took. It will all be reflected in his overt behavior and in his work. Usually professionals don't confess their negatives. They go straight ahead as though nothing is wrong. They murder their consciences, and thereafter commit such atrocities against humanity. Nevertheless, God, the righteous and just judge, is watching over all. One's heart can be transformed only through accepting Jesus as Lord and Savior.

Jesus told Nicodemus, "You must be born again—born of the Spirit." Godly parents are the children's role model. God says to the children in Exod. 20:12: "Honor thy father and thy mother; that thy days may be long upon the earth which the Lord thy God giveth thee." In conclusion the helpmate of the head of the house should always remember the words of God found in Eph. 5:22, "Wives submit yourselves unto your own husbands as unto the Lord; for the husband is the head of the wife even as Christ is the head of the Church; and he is the savior of the body." On the other hand, the husband needs to remember to love his wife as Christ loves the church. Both parents should also display love for their children, but should not "spare the rod" when the "rod" is needed.

The School

Schools are educational institutions. If the schools honor God, he will honor them and society. God says, "I will honor those who honor me." During the school season, most of the child's day is spent in school. Therefore, it is vital that parents know what their child is being taught and by whom. After all the parents are the taxpayers and Christian parents must not sit idly by and let the immoral segment of society impose its subversive lifestyle and immorality upon them and their children. Satan only needs to capture one generation to cause collapse and chaos in any society. In the United States, and other parts of the world, the devil is using immorality in the schools, in the media, in some homes, and in some apostate churches to cause societal collapse.

Academic excellence does not make one a moral person; neither does a man's great genius make him a man of wisdom. My friend, wisdom comes from God alone. God says, "There is a way that seems right to a man but the end is destruction thereof. Let the wicked forsake his way and the unrighteous man his thoughts" (Isa. 55:7). If the children were taught righteousness and morality in our schools as well as at home, there wouldn't be so much crime and violence there. Unfortunately, Satan has many anti-God, anti-Christ government agents running the system, and the righteous are currently playing a passive role, for the most part. Let this book wake you up, America, and get busy for the Lord our God. It's harvest time! Let's go to work! Be bold! Be strong! For the Lord thy God is with thee!

The Society at Large

In Ps. 124:1 the Lord says, "The earth is the Lord's and the fullness thereof, the world and they that dwell therein." This means that the entire society belongs to the Lord. Since Satan has infiltrated it through the sins of Adam and Eve, it is now necessary to have spiritual military bases in all points as a witness for Christ, and in order to rescue lost men who are in route to eternal destruction. God is concerned about human ecology, which is the branch of society concerned with the spacing of people and institutions and the resulting interpendency. We're not in this world alone. We depend on each other. Man

Basic Organization of the Local Church

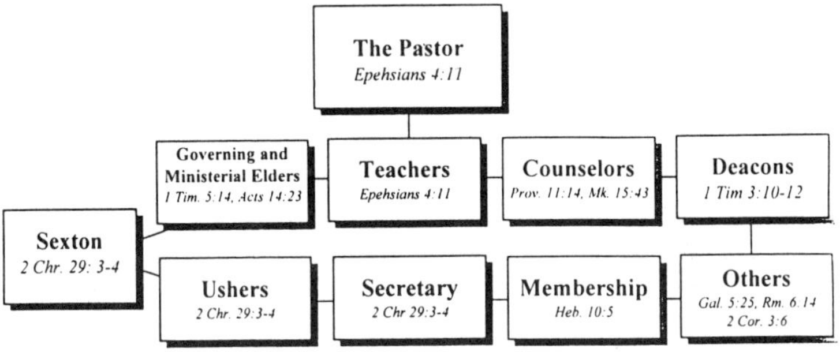

needs to learn the golden rule: "Do unto others as you would have them do unto you." When he learns and practices it, the society at large will become a better place in which to live. The golden rule will also have a tremendous impact on the behavior of one in his ecological niche, which is the position that a creature occupies in relation to his environment. Jesus says, "Come unto me all ye that labor and are heavy laden and I will give you rest" (Matt. 10:28).

In 1 Pet. 2:17, the Lord says, "Honor all men. Love the brotherhood. Fear God. Honor the King." If human beings applied this Scripture in our society, we would have a more peaceful one. Many social problems would vanish and we would witness the healing of a nation. The Word says the kingdom of God is within you. Through the leading and empowerment of God's Holy Spirit, we can tap into that zone of eternal power within. Now the power belongs to God and we can do nothing without His divine grace.

The Church

If a man honored the Lord at home, at school, on his job, and in all other areas of society, it would have a mighty impact on the praise and worship services conducted in his churches. Though the church that I shall write most about in this chapter is the physical structure and organization, I see the operation of another church in the spirit—and that is the global church of spirit-filled believers. No building can accommodate such throngs of people. Take a look at the map to your left and you will see the basic organization of the local church. The local church is an important spiritual military base. It's a place for the assembly of the local body of Christ. Pastors go there to feed the Lord's sheep. In Heb. 10:25 God tells us not to forsake the assembling of ourselves together. Therefore, the saints of God go to church to hear the preaching of the Word, to take the unsaved that they too might hear and give their lives to Christ, and for fellowship with fellow believers. A local church has to be organized and established as the Holy Spirit leads its founder. However, some traditional basic offices of a local church are:(1) the pastorate, (2) governing and ministerial elders, (3) teachers, (4) counselors, (5) deacons, (6) ushers, (7) secretary, (8) sexton, (9) membership, and (10) et ceteras. Let me give you a brief explanation of each office

The pastor. The pastor is the head of the local church. Based on Eph. 4:11 the Holy Ghost has placed him in the Body of Christ for ministry and for edifying of the body of Christ. He's God's basic mouthpiece in the local church.

Governing and ministering elders. 1 Tim. 5:17 says, "Let the elders that rule well be counted worthy of double honor, especially they who labor in the Word and doctrine." From this passage we can clearly see that there are two types of elders, namely governing or ruling elders and ministering elders. The governing elders assist the pastor solely with church government matters; elders who labor in the word and doctrine are ministering elders and they assist the pastor with ministry. For example, when the pastor is absent, they can easily serve as a substitute. They pinch-hit the ball. They're the pinch hitters.

Teachers. Based on Eph. 4:11, teachers are put in the Body of Christ for the building up and perfecting of the saints. In Tim. 2:15, God says, "Study to shew thyself approved unto God a workman that needeth not to be ashamed rightly dividing the Word of Truth." The Lord has put spirit-filled teachers in the Body of Christ to assist with our study of the Word. That's why we have Bible schools. However, I must warn you that those who are not called to teach should not teach because the Lord's anointing will not be on their teaching and they will be held accountable by God for their heresy. It is not who man anoints and approves who gets the job done. The anointing of God breaks the yoke.

Counselors. Prov. 11:14 says: "Where no counsel is the people fall: but in the multitude of counselors there is safety." Therefore, counselors are in the local church to encourage, advise and edify the sheep in need. They are a safety net for the sheep because by the grace of God they keep the sheep from disastrous falls. Counselors must be spirit-filled mature Christians who are experts in the Word of God, who sincerely love all people, and who can easily gain another's trust.

Deacons. Based on 1 Tim. 3:10–12, the deacons are men and women who have been tried and proven. They are holy men of God who walk uprightly. They are men who are concerned with the total church program. They love the sheep. Their job description is given to them by the local pastor of the church.

Ushers. Ushers are church helpers in the Body of Christ. They must be spirit filled, mature, and possess a wonderful personality because they set the tone of the worship service. The visitors see the ushers before they meet the pastor and elders. Though the ushers are involved in a ministry of helps, they have a key role at the church. They are like the Levites, doorkeepers in the house of God (2 Chron. 29:3–4). Ushers can cause church decline and they can also cause church growth. It all depends on the manner in which they perform their tasks.

Secretary. The secretary is the office keeper. She helps to keep the church in order through her excellent record keeping. She files, types, answers the phones, greets the guests of the pastor, arranges the pastor's appointments, etc. She must be spirit filled and have a pleasing personality. Her ministry is the ministry of help. She is like a Levite (2 Chron. 29:3–4).

Sexton. The sexton is the caretaker of the temple of God. He cleans the temple; he rings the bell if needed. He does janitorial work. His job description is given to him by the pastor. Sextons are also like the Levites (2 Chron. 29:3–4).

Membership. Heb. 10:25 tells the church members not to forsake the assembly of themselves together. Without membership there can be no church, and the leadership of the church needs to understand this. The members are precious in the sight of God. They are His sheep. The leadership is called to feed them. The Lord Jesus said unto Peter that day, "Feed my sheep." The sheep and the meadows belong to God. Without the membership there would be no church. The membership support the work of the Lord through their tithes and offerings and through their dedicated services unto the Lord. Members play a vital role in the church.

Et ceteras. Gal. 5:25 says: "If we live in the spirit, let us also walk in the spirit." Second Cor. 3:6 says, "The letter killeth but the Spirit giveth life." The Lord is telling us here that the servants of the most high God must be led by the Spirit and not by tradition. The pastor must add church positions as the Holy Spirit leads him. God has given him spirit-filled elders of great wisdom to advise him and to test the Spirit to see if it be of God. Elders should not be zero place holders, for they should be men of caliber whose services are utilized. Let the Holy Spirit lead you and you'll create other church positions and employment for the body if the need arises. We are living in the Dispensation of Grace. We have a better covenant. When a leader walks in the spirit and is led by Him, he is not under the Law. God is Spirit and grace causes the leader to know the will of God, and enables him to accomplish it.

Spiritual Weaponry

The Word of the Most High God is the Sword of the Spirit. (Eph. 6:17, Heb. 4:12)

IV

Spiritual Weaponry of a Christian Leader

Before he assumes a leadership role in Christian ministry, a leader must prepare himself spiritually, psychologically, academically, and physically. Spiritual preparation is necessary in order to know how to intelligently combat the adversary of God and man; psychological preparation will enable the leader to rest his mind when he becomes "weary and heavy laden." The Lord Jesus says, "Come unto me all ye that labor and are heavy laden and I will give you rest" (Matt. 11:28). "There remaineth therefore a rest to the people of God. For he that is entered into His rest, he also has ceased from his own works, as God did from His" Heb. 4:9–10.

God is the psychologist of psychologists. He understands the human mind. For He created it and only He can provide rest or peace to the weary. Academic preparation (whether it is obtained from the school of life's experience or from a cultural and religious institution of higher learning) is extremely vital. The Old Testament teaches that God's people perish for lack of knowledge. Yes, it is true that God is not looking for ability but availability, because He teaches and equips all whom He chooses for ministry.

It is also true that God puts no premium on ignorance. Ignorance is a tragedy indeed! Because of the tremendous demands of a leadership role in ministry, the leader must be very knowledgeable, resourceful, and well acquainted with spiritual military techniques of war. Physical preparation is necessary because the Bible teaches that the body is the Temple of the Holy Ghost and God does craftsmanship within man. Generally speaking, in order for God to use a leader widely and effectively, the leader should be physically fit. A tired, worn-out person cannot provide dynamic Christian leadership services for the

Kingdom of God. Therefore, the leader should eat nutritious food, and exercise often, if at all possible. Diet experts say proper food can have a positive effect on one's emotions, but improper nutrition can have a negative effect on human emotions. The dieting observation is a generalization. Remember, that for every generalization, there are exceptions.

In the introduction of this chapter, we learned that a leader needs to be well acquainted with spiritual military techniques. However, knowledge alone of spiritual military techniques cannot win our spiritual battles in high places. With this truth in mind, we are going to focus our attention on the spiritual weaponry of a Christian leader in subsequent paragraphs of this chapter.

The Sword of the Spirit

The Holy Bible teaches that the Word of God is the sword of the Spirit. To quench the fiery darts of the wicked, the Lord says in the New Testament, "and take the helmet of salvation and the sword of the spirit which is the Word of God" (Eph. 6:17). In the Old Testament, the Lord through King David says, "Thy Word is a Lamp unto my path" (Ps. 119:105). Conclusively, we can say the Word of God is a light unto our paths and it is unquestionably the sword of the Spirit. It is the indisputable sword of the Spirit because from the Word we learn how to defend ourselves against our scheming adversary, the devil. "For the Word of God is quick, and powerful, and sharper than any two edged sword, piercing even to the dividing asunder of soul and spirit, and of the joints and marrow, and is a discerner of the thoughts and intents of the heart" (Heb. 4:12).

The Whole Armour of God

Victory or victimization? You may elect to be a victor or a victim. The choice is yours. My friend, you're on the battlefield of the Lord and your adversary, the devil, has declared war against you. Therefore, it is absolutely necessary that you put on the whole armour of God, which will enable you to withstand the enemy in the evil day. The New Testament tells us that the armour of God consists of truth, the breastplate of righteousness, feet shod with the preparation of the Gospel of Peace, the shield of faith, the helmet of salvation, the sword of the spirit, constant prayer and supplication in the spirit and watching with all perseverance and supplication for all saints. If you choose to put on the whole armour of God, you're destined for victory. But if you choose not to put on the whole armour of God, your destiny will be victimization.

The Word of God says, "For though we walk in the flesh we do not war after the flesh: (for the weapons of our warfare are not carnal, but mighty through God to the pulling down of strongholds;) casting down imaginations and every high thing that exalteth itself against the knowledge of God, and bringing into captivity every thought to the obedience of Christ" (2 Cor. 10:3–5). According to 1 Sam. 17:45, the battle is not ours. For the battle is the Lord's. Putting on the whole armour of God means that you have followed the Lord's guidelines for spiritual warfare and He will surely honor his Word and give you the sweet victory. The Lord is strong and mighty in battle. Clothed in Christ's righteousness, the leader must say like King David, "I will lift mine eyes to the hills from whence cometh my help" (Ps. 121:1). Meditate on the knowledge of the power of God that is within you, use the spiritual sword with expertise, and I guarantee you the Christian flag will be exalted after each battle. Remember that prayer, fasting and the giving of alms are three of the most effective weapons that a Christian can use against the endangered species (Satan and his fallen angels).

Plans and Strategies

V
Plans and Strategies

A leader without a plan is like a constructor of an architectural structure without a blueprint. Without a plan of action, how can a leader intelligently direct his followers? Without a blueprint, how can the chief constructor instruct his construction workers? Plans and strategies in leadership are absolutely indispensable. When a leader fails to plan he is simultaneously creating an invisible plan called "Failure." Failing to plan results into overwhelming failure! Jesus said in the New Testament, "For which of you, intending to build a tower, sitteth not down first, and counteth the cost, whether he have sufficient to finish it" (Luke 14:28).

In that passage the Lord Jesus is addressing the financial aspect of construction. However, the addressing of needed finances would be inescapable in the leader's detailed outline or plan. Now the question before us all is, "How do we spell success in leadership?" Let me hasten to give you the answer. A leader's detailed plan under the anointing of God's Holy Ghost spells success in leadership if the anointed leader follows the plan. Planning is thought provoking and time consuming. Planning is **work** within itself! Always keep in mind that the only place you will find that **"success"** comes **before** "work" is in the dictionary.

Now let's examine leadership plans and strategies from a good old Christian perspective. What is a plan? What is a strategy? How are plans implemented? What results do they produce? These and other questions that are perhaps in your mind will be answered in this chapter.

Plan

A leader's detailed outline of an assignment inspired by the

Holy Spirit of God. The outline must be one of clarity that tells us what, when, where, how, how much, etc. It must answer the following questions: What is the plan about? What are my goals or objectives? Where am I going? How am I going to get there? How will I know when I have arrived?

Strategy

A strategy is a method of obtaining a specific goal. For strategies the leader must seek the mind of God, and the Holy Spirit of the Lord will reveal to him the strategies that will lead him to victory. Sometimes it will be necessary to fast in order to acquire directions for the application of proper strategies. Remember Ezra? The Old Testament teaches that he and the Levites planned a journey to Jerusalem to take silver and gold and other substances to the House of the Lord. Knowing that there would be enemies along the path, Ezra proclaimed a fast. The Word tells us that he was ashamed to secure the soldiers and horsemen as help from the king against the enemy in the way because they had spoken to the king saying, "The hand of our God is upon all them for good that seek him; but his power and his wrath is against all of them that forsake him." Ezra and the Levites took it to the Lord. He says, "So we fasted and besought our God for this and He was entreated of us" (Ezra 8:23). The prayer and fasting strategy worked because it was orchestrated by God. The Lord delivered Ezra and his people from the hands of the enemies.

The Lord of Hosts is a deliverer. God has a divine plan for everything that He assigns us to do. Seek him and he will reveal the human strategy to you. Furthermore, the Lord will give you "a peace which passes all understanding." You're just the human vessel that God uses. Let him do the planning and directing. "Love never faileth." That's why the Lord says, "In all thy ways acknowledge him, and he shall direct thy path" (Prov. 3:6). The Bible also teaches us that those who labor to build a house without the help of God labor in vain. To succeed your strategies must be inspired by the Lord who is mighty in battle.

Considering All Possibilities and Probabilities

When a leader plans an attack against his enemies, he must consider possibilities and probabilities in the confrontation. Only then can he plan intelligent strategies to combat his foes. We wrestle against principalities, powers, the rulers of the darkness of this world, and spiritual wickedness in high places according to the Word of the Lord. The Most High God says in Jer. 33:3, "Call unto me, and I will answer thee and shew thee great and mighty things which thou knowest not." Successful strategies come from God. The battle is not ours, for it belongs to God, the Almighty. Inasmuch as we are the craftsmanship of God, we must ask ourselves the following question: "What king going to make war against another king, sitteth not down first and consulteth whether he be able with ten thousand to meet him that cometh against him with twenty thousand" (Luke 14:31). Consider the possibilities and probabilities. Then make your judgement based on your findings. What a strategy! And what a victory! Victory is yours indeed!

To close this chapter, I want you to take a look at the famous strategy used by Joshua after the Lord told him, "I have given into thy hand the King of A'i, and his people, and his city, and his land" (Josh. 8:1). I also want you to take a look at the successful strategy of Hushai against Absalom, King David's son, who wanted to annihilate his father and take over the throne.

Joshua's Strategy against King A'i and His City

The Old Testament teaches that the Lord told Joshua to fear not because he had given into his hand King A'i and his city. With faith in the Word of the Lord, Joshua planned the following strategy: he chose thirty thousand mighty men of valor and sent them away by night. He commanded them, saying, "Behold ye shall lie in wait against the city, even behind the city: go not very far from the city but be ye all ready. And I, and all the people that are with me will approach unto the city: and it shall come to pass, when they come out against us, as at the first, that we will flee before them. (For they will come out after us) till we have drawn them from the city; for they will say they flee before us, as at the first. Therefore, we will flee before them. Then ye shall rise up from the ambush and seize upon the city; for the Lord your God shall deliver it into your hand" (Josh. 8:4–7). Joshua's strategy worked because the Lord God ordained it. Furthermore, the most High God went to the battlefield before Joshua and his army. He prepared the way for his servant. The New Testament teaches "If God be for me, who can be against me." No weapon formed against you will prosper when you walk humbly with God.

Hushai's Successful Strategy against Absalom

The Scriptures tell us Absalom led a rebellion against his father, King David. Wanting rulership, the throne being his desire, Absalom literally tried to destroy the king. Therefore, he sought counsel from Ahitophel and Hushai. Now Ahitophel's counsel could have easily defeated King David had Absalom listened to it. Instead he sought further counsel from Hushai. And Hushai convinced him not to follow Ahitophel's counsel. The strategy was successful because the Lord was with Hushai. In the Word we are told, "For the Lord had appointed [Hushai] to defeat the good counsel of Ahitophel, to the intent that the Lord might bring evil upon Absalom" (2 Sam. 17:14).

I have used the military strategies of Joshua and Hushai to bring home a spiritual point again: the point is as leaders we are in the army of the Lord. You may never join the military of your country, but if you accept a leadership role from God, you are automatically joining God's army. You will constantly have to seek God's counsel because you will need to develop plans and strategies for your work. Success is yours because you have put all your trust in King Jesus. However, you must always remember that "faith without works is dead" (James 2:17).

The Secret Place

VI
The Secret Place

Every creature of God has a secret place, which was built within him during his creation. What is it and what is the purpose of it? I shan't keep you in suspense. Let me quickly answer your twofold question. A man's **heart** is his secret place. His eyes are only the windows of his soul. The heart is quite revealing! It tells us what a man is thinking, what he feels, and what he believes. Only the Creator has the power to enter man's secret place to find out his innermost secrets. God is the **heart specialist.** The Word of God teaches, "For a man looketh on the outward appearance, but the Lord looketh on the heart" (1 Sam. 16:7).

One can easily deceive man but not God. Our omniscient God knows the intents of the heart. Nevertheless, when He created man, He gave him a heart, and constructed a door with a lock on it at the entrance of the heart. Knowing that he had also given man a free will, he decided to give the key to man. Now we can easily understand why the Lord Jesus says in Rev. 3:20, "Behold I stand at the door and knock. If any man hears my voice and opens the door for me, I will come into him and eat with him and he with me."

The Lord is a gentleman. He is not an intruder. He ***knocks*** at the doors of men's hearts. If they respond to his knocking, he enters in as he said he would. If they do not respond, the Bible teaches that God's spirit will not always strive with man. The Lord will turn them over to a reprobate mind. ***We*** need the Lord! ***He*** does not need ***us.*** However, He ***delights*** in using ***obedient*** servants as His vessels of honor.

Letting Jesus into your secret chambers means giving the rulership of the throne of our heart to him. You are no longer ruler, and in his infinite wisdom, the Lord will not share the throne with anyone. He alone must be your Lord. You are a temple of the Holy Ghost. As a leader you must not let anybody enter into the secret place of your temple. That means Mother can't enter it, Father can't enter it, wife or husband can't enter it, nor can the children enter it. If you put Jesus first, he will assure you that no one else will enter the secret place. Jesus, the Lion of the Tribe of Judah, sits on your throne and watches over you. But to the disobedient who allow their loved ones to enter the secret place, Jesus says, "Think not that I am come to send peace on earth; I came not to send peace but a sword. And a man's foes shall be they of his own household. He that loveth father or mother more than me is not worthy of me; and he that loveth son or daughter more than me is not worthy of me" (Matt. 10: 34, 36–37).

Your obedience to God will avert His usage of the sword. Take up your cross and follow Jesus. Say to your loved ones, "Only Jesus is permitted to enter the secret place." Then you will be a leader loved by God. Moreover, your loved ones and friends will know that your walk with the Lord is reality not make-believe. Say inaudibly, "I can do all things through Christ which strengtheneth me" (Phil. 4:13).

A Sound Counsel to All Readers

My advice to you is simple, down to earth, and straight from the throne room above. If you have inadvertently allowed someone to enter your secret place, I advise you to speedily escort the person out. Let him know that no visitors are allowed in the secret place. Jesus dwells there alone.

Benefits of the Reign of Christ in Your Secret Place

When Christ reigns in the secret place, the leader or believer dwells in the shadows of the Almighty. The Lord upholds him with his right hand of righteousness. The Lord will cover him with his feathers, and under his wings shall he trust. God's truth shall be his shield and buckler. "He that dwelleth in the secret place of the Most High shall abide under the shadow of the Almighty. I will say of the Lord, He is my refuge, and my fortress: my God; in Him will I trust" (Ps. 91:1–2).

Let us pray: "Lord, thou hast been our dwelling place in all generations. Before the mountains were brought forth, before you formed the earth and the world you were God. We trust your infinite wisdom. Remember our frailties. For we are just a lump of clay formed out of the dust of the earth. Without thee, Almighty God, we can do nothing. Father, we commit our secret place unto you that you might reign therein forever, and ever. Reign in us, sovereign Lord, reign in us. Create within your servants a heart of righteousness. Enable us to walk in holiness redeeming the times. Teach us to number our days that we may apply ourselves to wisdom. This we ask in the Holy name of Jesus, thy son. Amen."

Spiritual Laws to Remember

Law of Love
(Mat. 22:37-38, 1Cor. 13, 1 Tim. 6:10, Jn. 12:35)

Law of Sowing
(Hos. 8:7, Job 4:8, 11Cor. 9:6)
(Corn=Corn, Rice=Rice)

Law of Transformation
(Rom. 12:2)

Law of Restraint
(Ecc. 2:13, Ex. 18:12'27, 2Pet 3:18,
Phil. 4:6, Prov. 19:23)

Law of Exaltation
(Luke 18:10-14, Jas. 4:6, Acts 2:22, 2 Cor. 18:18)

Law of Reciprocity
(Luke 6:38)

Law of No Return
(Heb. 6:4-6)

Law of Loveless Gaining
(Mat. 16:26, Prov. 16:32, 11 Tim. 4:10)

Law of the Church
(Mt. 7:16;12:30)

Law of Subtraction
(Mt. 25:14-30)

Law of Multiplication
(Mt. 25:14-30)

Law of Spiritual Warfare
(11 Tim. 2:4, 1 Tim. 1:18'19)

Law of Creation
(Mal. 2:10, Acts 17:26, Gen. 1:26,
Gal. 3:28-29, Acts 10:34)

VII
Spiritual Laws to Remember

Fellow servants, there are spiritual laws in the universe to be observed and utilized to the highest degree possible. Failing to apply them will result in Christian leadership failure. But application of them will result into victorious Christian living. Just as there are physical laws in our universe, there are spiritual laws. Perhaps you might argue, "We're under the dispensation of *grace*. We're not under the dispensation of *law*." If that's your argument, I have good news for you from the *Good News Holy Bible*. The Lord says, "For sin shall not have dominion over you; for ye are not under the law but under grace. What then? Shall we sin because we are not under the law, but under grace? God forbid" (Rom. 6:14–15).

God's grace is a better covenant than the covenant of law. He expects more from those of us living in this dispensation of grace. The Lord now requires all men to repent. The Holy Spirit has come now, and through His divine assistance, we can walk in victory. We walk in the spirit and are Christ conscious. That's why we neither disobey the Law nor think it's all right to ignore it. On the contrary the Apostle Paul said, "Walk in the Spirit and ye shall not fulfill the lust of the flesh" (Gal. 5:16). According to the New Testament, the Law is fulfilled in one word. And the word is *love*. We should love our neighbors as ourselves. All spiritual laws hinge on God's divine love.

Then and Now

God had set physical laws in the universe before He gave the prophet Moses His divine laws on Mount Sinai. The physical laws that existed before the divine laws are still in existence today. A good example is from the Laws of Motion. Let's elaborate briefly on God's first law of motion. The first law of motion is the law that a body remains at rest or in motion with a constant velocity unless an external force acts on the body. It is the law that the sum of the forces acting on a body is equal to the product of the mass of the body, and the acceleration produced by the forces with motion in the direction of the resultant of the forces. This is one of the three laws of classical mechanics. It's physics. Our God is a physicist. He knows all about physics. He created it. Now the law is called Newton's Law of Motion, but I have news for you. The law of motion has been in motion since the time of creation. God put that law in motion because all things were made by him. Isaac Newton was only allowed by God to discover the law. Let human discoverers decrease and let the Lord increase in their lives. Let them give God the glory.

Recently, I went to Ghana, West Africa, to conduct a Christian leadership seminar before many Ghanaian pastors. It was held at the New Testament Assembly in Kumasi. Obviously, in preparation for the seminar, I sought the mind of God for content. I did as Ezra said he did: "I sought the mind of God with prayer and fasting and he was entreated of me" (Ezra 8:23). The Most High God inclined His ears to hear His servant. One of the items He told me to include in the seminar was the subject of spiritual laws to remember. As the Holy Spirit escorted me through the Old and New Testament, He showed me thirteen laws that he wanted me to include in the seminar. In obedience I included them and we all had a marvelous time at the pastors' seminar in Africa.

Before I give you the spiritual laws to remember, let us ask, "What is law and to whom is it applicable?" Law is a principle established by a government and applicable to a people. There are divine laws, governmental laws, and scientific laws. The laws that we shall examine in the succeeding paragraphs of this chapter are the Lord's spiritual laws. They come from our divine government above. God is reigning even today over spiritual Israel, and we are seated with him in high places, spiritually speaking.

Law of Love

Love God with all your heart, with all your soul, with all your mind, and love thy neighbor as thyself. This law of love is based on Matt. 22:37, 38.

The Lord tells us that love covers a multitude of sins. Moreover, love is a fruit of the Holy Spirit. Only a spirit-filled believer can display genuine love. Our God is the reigning champion of love. He displayed his love when He gave His only begotten son to the world as the sacrifice. "Greater love hath no man than this, that a man lay down His life for His friends" (John 15:13). Love is compassion. It means sharing and caring. If we love God we will not practice sinning against Him. His Holy Spirit enables man to live a life of holiness. Love is the key to all spiritual laws.

Law of Sowing

Man reaps just what he sows. "He which soweth sparingly shall also reap sparingly; and he which soweth bountifully shall reap also bountifully." If men sow "the wind they will reap the whirlwind." If one plants corn in his field, he will get corn at harvest time. If he plants tomatoes, he will get tomatoes in the harvest. One does not plant beans and expect mustard greens in the harvest. The Law of Sowing is based on 2 Cor. 9:6, and Hos. 8:7.

Law of Transformation

Be ye transformed by the renewing of your mind that ye may prove what is that good, and acceptable, and perfect will of God (Rom. 12:2). An unregenerated person who rejects the love of the Lord will never know the good, acceptable, and perfect will of God. Only the transformed will learn God's will. Based on the Old Testament, a man is conceived in sin and shapen in iniquity. He needs to be transformed before he can develop an intimate relationship with God. As noted, this law of transformation is based on Rom. 12:2.

Law of Restraint

"Be ye angry but sin not; let not the sun go down upon your wrath. Neither give place to the devil" (Eph. 4:26–27). Zeal without wisdom can lead to folly. The Word says, "Wisdom excelleth folly as far as light excelleth darkness" (Eccles. 2:13). God has given us a fruit of the spirit called temperance or self-control. Use it. It enables one to restrain himself whenever necessary. If one fails to apply the Law of Restraint, he will "surely wear away" as Jethro tells Moses in the Book of Exodus. God uses restraint, because He is longsuffering. The Word says, "The Lord is angry with the wicked all day long." God uses restraint because He sent His Son, Jesus, into the world not to condemn it but that the world through Him might be saved.

Law of Exaltation

"Everyone that exalteth himself shall be abased; and he that humbleth himself shall be exalted" (Luke 18:14). So if you want to be exalteth, humble yourself before the Most High God. The Lord resists the proud, but He gives grace to the humble. Decrease. Let your light shine before the sheep and the lost that they might see Jesus in you. Let Jesus increase His stature in you and you will automatically be exalted.

Law of Reciprocity

"Give and it shall be given unto you; good measure, pressed down, and shaken together, and running over, shall men give into your bosom. For with the same measure that ye mete withal it shall be measured to you again" (Luke 6:38).

Fellow servants, let me give you an omen. I must warn you

that the Law of Reciprocity does not refer solely to finances. Instead it refers to God's abundant provisions. God is concerned about all of man's needs. For He invests in His total program for man.

Law of No Return

"For it is impossible for those who were once enlightened, and have tasted of the heavenly gift, and were made partakers of the Holy Ghost, and have tasted the good Word of God, and the powers of the world to come, if they shall fall away to renew them again unto repentance, seeing they crucify to themselves the Son of God afresh, and put Him to an open shame" (Heb. 6:4–6). The Law of No Return is based on the Gospel of Jesus Christ. It is based on truth. Let no one deceive you with the myth "once saved always saved." It is possible, but it's not necessarily so. How about King Saul? And how about Lucifer, the anointed of the Cherubim Angels? King Saul had a good beginning, but he failed to surrender himself totally to God. Scriptures say he sought repentance with tears but found no room for repentance. The Lord resisted King Saul because of his disobedience and arrogance. Saul committed suicide.

Lucifer was the worship leader in Heaven. He protected the throne of God, and he was the anointed Cherub. Unfortunately, he corrupted his wisdom because of his beauty, the Scriptures tell us. It is true that if you surrender yourself completely to God, no one can pluck you out of his hands. However, that is contingent upon your covenant relationship with the Lord.

Law of Loveless Gaining

"For what is a man profiteth if he shall gain the whole world and lose his own soul? Or what shall a man give in exchange for his soul?" (Matt. 16:26).

All good gifts come from the Father of Lights. You don't want anything that doesn't come from God. I advise you not to accept anything that comes from a source other than God. To accept would mean *loveless* gaining. You gained it through evil means, and if you continue in the path of loveless gaining, your soul will be required of you. What will a man give in exchange for his soul?

Law of the Church: For Me or Against Me

"He that is not with me is against me; and he that gathereth not with me scattereth abroad" (Matt. 12:30). To be a member of the Body of Christ, which is the church of the Lord, one must be for him or on the Lord's side. By their fruits they are known. There is no place for neutrality in this world. Jesus says you are either for me or against me. Neutrality is a deception of Satan. It is impossible to live a neutral life in this world. You have to serve somebody. If you serve the Lord, you will say so. But if you serve Satan, you don't have to say so or agree with the fact. God's truth marches on! "Let the redeemed of the Lord say so."

Law of Subtraction

Failing to use a God-given talent will result into His taking it away and giving it to a servant who owns more. He gives it to the servant who has more because that servant has demonstrated gratitude and servantship. The useful servant is productive and pleases the Master, but the slothful servant brings his Master displeasure. It all results into the Law of Subtraction. This is based on parable of the talents (Matt. 25:14–30).

Law of Multiplication

From him whom much is given, much is required. When the Lord gives a servant much, he requires much production. The more he gives you, the more he requires. If you are faithful, he will multiply your blessings. The Law of Multiplication is also based on the Parable of the Talents (Matt. 25:14–30).

Law of Spiritual Warfare

"No man that warreth entangleth himself with the affairs of this life, that he may please him who hath chosen him to be a soldier" (1 Tim. 2:4). As a servant of the Lord, just keep your eyes focused on Jesus, the bright and morning star. He will guide you through the Holy Ghost. Fight the war with a clear conscience, and you will surely be more than a conqueror in Christ Jesus. The victory will be given to you because the battle is the Lord's.

Law of Creation

God has "made of one blood all nations of men for to dwell on all the face of the earth, and hath determined the times before appointed, and the bounds of their habitation." "Have we not all one father? Hath not one God created us?" (Acts 17:26 and Mal. 2:10).

Out of one blood, God has made all men. And out of one blood, he redeems all men. From Adam and Eve, the black men descended, the white men descended, the brown men descended, the red men descended, and the yellow men descended. The different races of people are simply artistic expressions of God's divine imagination and of God's divine creativity. Our God is a God of variety and creativity. If you take a good look at all of His creation, you'll find variety. There is variety in his trees, his animals, his fowls, his birds, his flowers. This same God has created a variety of human beings, and they are all precious in his sight. They are like a bouquet of flowers. God tells us in Acts 10:34 that he is no respecter of persons. In other words, He is not a prejudiced God. He's a God of Love. God tells us in the New Testament that heaven will be filled with people from every nation, every kindred, and every tongue. If a leader is not free of prejudice, he should not be trusted to rule over someone whom he hates without a cause. He should concentrate on developing a wholesome relationship with God and not leadership because he cannot be trusted by all people. His feelings and attitude will be reflected in his leadership. God desires spirit-filled leaders, not divisive ones.

In closing this chapter, I am going to give you a brief description of creation of man as told in the Book of Genesis: "So God created man is His own image. In the image of God created He him. Male and female created He them" (Gen. 1:27). In other words, God created Adam first. While Adam was asleep, the Lord took one of his ribs and made Eve. The Lord commanded Adam and Eve to replenish the earth, to be fruitful and multiply. Therefore, using Adam and Eve as reproductive vessels, the Lord continued His creation of all mankind. Let us celebrate the physical differences. Why not think about a man's character? On Judgement Day, God will judge a man's character, not his physical features or race. Those things are irrelevant to God. Whenever a man surrenders himself to God, the latter aspects also become irrelevant to him. God does not change a man's race or color because he knows that a man's sins are not hidden in his ethnicity. For they are hidden in his heart. Therefore, God changes only the heart when a man surrenders. For "the heart (not skin tone or color) is desperately wicked," saith the Lord of Hosts.

Leadership Urgency

VIII
Leadership Urgency

The Word of God teaches that the coming of the Lord will be like a thief in the night. No one knows the exact day or hour, but based on Bible prophecy and the signs of the times, our redemption draweth nigh. Therefore, it behooves all of God's servants to get busy in the Master's vineyard while it is still day. Night will come and no man will be able to work. I urge every pastor, apostle, prophet, evangelist, teacher, and layman to put on his sword and shield. For it is time for war, not retirement or resignation. There is no retirement in God's army. For the Lord said, "Be thou faithful until death, and I will give thee a crown of life" (Rev. 2:10).

On the other hand there is no resignation in God's army. The Lord also said, "Let us run with patience the race that is set before us" (Heb. 12:1). Our holy and righteous God will not reward us for beginning the race that He set before us, but He will reward us for finishing the race. The race is not over until it's over. In other words your race is not over until the Lord says so, for "the race does not belong to the swift nor the battle to the strong."

What is the urgency in leadership? And what can we do about it? The urgency in leadership is harvesting the crops in the fields of the Lord. It's soul winning. Put your hands to that sickle and swing, swing, swing! Only then can we reap the harvest.

The Holy Bible teaches us three absolute truths about the harvest. They are: (1) the great harvest truth, (2) the ready fields for harvest truth, and (3) the harvest is past and the summer is ended truth.

Before we plan our strategies for gathering the harvest, let us examine the three absolutes mentioned above. Meditation on the absolutes first will enable you to seek directions from the

Holy Spirit more intelligently as to how to gather the food (souls) for the harvest.

The Harvest Truth

Luke 10:2 teaches us, "The harvest is truly great, but the labourers are few." And what is the solution? The Lord tells us in the same passage to pray therefore to the Lord of the harvest that he would send forth labourers into his harvest. In this absolute we observe that the harvest is abundant, but it has only a few labourers. Since the true Christians know that the earth is about to witness the heavenly sunset, it is urgent that everyone gets busy for the Lord. To be successful, true born-again spirit-filled Christians must unite universally in a concerted effort to win souls for our Lord and Savior Jesus Christ. The sheep are going astray. There are lions and wolves among them. The shepherd boys are on the haystack soundly sleeping. Let the Lord sound the alarm and awake them before his sheep are devoured. Let the Lord awake the shepherds, his anointed ones. He whom the Master finds asleep when he comes will be left behind. Let us pray to the Lord of the Harvest as God commanded us, and he will surely send laborers to His harvest. The harvest belongs to the Lord. God says, "For every beast of the field is mine, and the cattle upon a thousand hills" (Ps. 50:10).

The Ready Fields for Harvest Truth

Jesus teaches us the following in John 4:35: "Say not ye, there are yet four months, and then cometh harvest? behold, I say unto you, lift up your eyes, and look on the fields; for they are white already to harvest." It doesn't take a telescope to discover facts in this Scripture. For the facts are all at our fingertips. Therefore, we can use our spiritual microscope for that which is at hand. We can easily see that the fields are white and ready for harvest. And we easily hear the uninterrupted sounds of the voices of the people of Macedonia crying for assistance. "Come over to Macedonia and help us," they cry. We can also hear the voices of the unsaved in America crying out to God for mercy. We, the servants of the Most High God, must respond to the calls from afar and nigh. From afar because God said, "Go ye into all the world and preach the Gospel." From nigh because the Lord said, "What does it profit a man if he gained the whole world and lost his own soul?" What would America gain or what would any country gain if it gained the whole world and lost its own soul? You answer that.

The Harvest Is Past and the Summer Is Ended Truth

In Jer. 8:20–22 the Lord tells us, "The harvest is past and the summer is ended, and we are not saved. For the hurt of the daughters of my people am I hurt. Is there no balm in Gilead? Is there no physician there?" If we meditate on this passage and begin or continue laboring faithfully in the fields, we will never have to confess to our Lord, "The harvest is past, the summer is ended, and we are not saved." We have a balm in Gilead! His name is Jesus, the Messiah! We have a physician in the vineyard. His name is Jesus, the great physician!

My friends, there is an urgency in leadership. That urgency is the salvation of the lost. "Now is the acceptable time. Today is the day of salvation." We have all the tools that we need. Let us go on the highways and byways of life, and tell men the Good News: Jesus died to save us all. Let's tell them about the love of the Lord. But let us not hesitate to tell the lost that the wrath of God falls on the children of disobedience. Be sure to tell them God's love covers a multitude of sins.

Stewardship of a Leader

IX
Stewardship of a Leader

God affirms in the Old and New Testaments that He owns the world and they who live within it are His creatures. Furthermore, the affirmant tells us that all things were made by Him and without Him nothing has ever been made. Without Him nothing will be made! Though God is the landlord of the world, He has assigned man the role of stewardship. We know that God is concerned about all human affairs. He cares about our governmental, our religious, educational, cultural, and social institutions. God cares about the total man and His environment. Unfortunately, the stewards of the world's institutions are not all godly men. Some ungodly men even preside over religious institutions. God hasn't called the ungodly to be stewards over any of His affairs. However, He has allowed their reign through His permissive will. "The wheat must grow with the tares until the harvest" (Matt. 13:30). In this chapter we are going to take a telescopic look at the stewardship role of a Christian leader. We are going to define steward and godly stewardship. Then we will look specifically at the duties or responsibilities of a good steward.

Steward

According to the dictionary, a *steward* is one who manages another's property or financial affairs. Spiritually speaking, God has chosen Christian leaders to manage His property and financial affairs in their homes, in their Christian schools, and in their churches. The Bible teaches us to be good stewards over that which the Lord gives us.

Stewardship

In a spiritual sense, stewardship is a trust granted for profitable use. God is a progressive God, and we are created in His image. Therefore when He gives a son or daughter stewardship over something, He expects to see action taken on the part of the steward. The Lord loves progressive stewards, but He hates slothful servants. In the parable of the talents, we learn that the wrath of God fell upon the slothful servant to whom the Lord gave one talent. The slothful servant took the talent and buried it. The Master told him that he could have taken that one talent to the money exchangers and made interest on it, so when the Master returned he would have been able to show him the interest. He failed God. So he was cast into outer darkness where there was weeping and gnashing of teeth. On the other hand, the other two servants to whom God gave two and five talents were very productive. The Master was well pleased with them. So he told them, "Well done thou good and faithful servant, thou has been faithful over a few things. I will make thee ruler over many things. Enter the joy of the Lord."

When a servant has done all that is possible in his stewardship role, he should then enter into his rest. After all, the fields, the meadows, and all the brooks of this life belong to the Lord. He only asks the leader to be a good steward over his meadows. As the leader relies on the Christ within, he will fulfill God's desire. God has designed success for His people not failure. "Christ in you is the hope of glory" (Col. 1:27).

Duties of a Good Steward

A good steward has many responsibilities, for a stewardship assignment is quite demanding. In the remaining paragraphs, I shall list some of a good steward's responsibilities. However, I want you to remember that it is necessary to be sensitive to the leading of the Holy Spirit in stewardship. He will guide you and show you all things. Some of the duties of a good steward are as follows: Discharging domestic duties (Luke 8:3), paying laborers a fair salary (Luke 16:1–13), providing wisely for those over whom he reigns (Luke 12:42); always being prepared to render account of his household or ministry (Matt. 18:23); being a productive steward (Matt. 21:34). As you develop your job, so will your stewardship duties increase. From him "to whom much is given much is required," says the Lord. The Holy Spirit gives gifts to the servants of God according to the will of the Lord (Matt. 14:10). When he bestows gifts upon us, he expects us to use them for the Master's glory and honor. God expects us to glorify Him in whatever office He gives us in the Body of Christ (Eph. 3:2:10) and (Titus 1:7).

The secret of stewardship is that the steward decreases and the Lord Jesus increases in His life. The steward learns to lean on Jesus. He recognizes that he has been entrusted with an office or ministry of the Lord. God wants to do a work through him. In closing this chapter on stewardship, I want you to remember that it is not production alone but it is also the quality of what one produces that counts. A combination of quality and quantity is ideal, but quality supersedes quantity. Jesus said, "I am the vine. Ye are the branches. Apart from me ye can do nothing." Seek the mind of the Lord for direction and be faithful on your job, because "faith without work is dead." Remember only the Lord can give the increase. For the meadows and the brooks belong to God.

Accountability to the Most High God

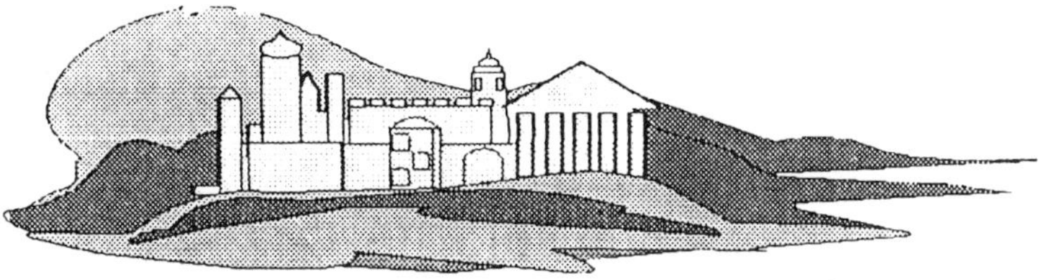

X
Accountability

The New Testament declares that we shall all stand before the judgment seat of Christ to account for our stewardship. How exciting! It is a day to which all true spirit-filled believers look forward. It shall be a day of rewards and honors for the faithful but a day of disappointments for the unfaithful.

Each person shall be judged by the work he did in the Body of Christ. That's going to be a glorious day because God has not programmed anybody for failure. On the contrary the Lord says that we are more than conquerors through Christ Jesus. The Lord has spoken quite clearly about His desire to reward His servants for their faithful servantship. True it is that the Lord has done the craftsmanship within the servants. True it is that all the glory and honor belong to God. That's why the Lord says in Matt. 5:16, "Let your light shine before men, that they may see your good works, and glorify your Father which is in heaven." At this point, I want to make it clear that the judgment seat of Christ in this chapter does not deal with the judgment of the wicked. Instead it deals with judgment and rewards for the righteous. It's the Bema Seat of Christ. Let no one think that the labor of love of the saints is for salvation. For no sinner can work to earn salvation. Christ's shed blood on the cross at Calvary earned salvation for those who believe in Him. Furthermore, the Lord said in Eph. 2:8–9, "For by grace are ye saved through faith, not of works. It is a gift of God. Not of yourselves lest any man should boast."

To the servants in the Body of Christ who are not giving God your best, I urge you to reconsider what you're doing. You began your labor of love with great zeal. So recharge your batteries and work with the zeal you had when you first knew him. Christ wants the best for you. To get His best for you, it will be necessary for you to give Him your best. God said, "I will honor those who honor me."

Throughout the Holy Bible, we read news about accountability. Some good news, and some bad news. The news is good when the servant has obeyed the Master and has been a good steward over that which the Lord has given him. On the other hand, the news is bad when the servant has been ruthless, slothful and disobedient. Be that as it may, in the subsequent paragraphs of this chapter, I am going to elaborate on four biblical accounts of accountability.

God Accounts for Man, His Creation

God alone is judge and he doesn't have to give account to anybody for anything. However, because of His nature, He is accountable to His conscience for everything He has created. He knows how many; He knows where; He knows when; and He knows how. God knows the count of the strings of hair on a man's head. The New Testament tells us, "But the very hairs of your head are all numbered" (Matt. 10:30). Accountability is divine and it is good. If one took an in-depth survey of any outstanding organization, one would find that accountability makes it tick. The director and employees work together. Those who do not perform according to job demands and expectation are released. If a steward cannot account for his stewardship, he should be released because with him on the job, the organization is headed toward the road of no progress. Let's take a look at Ps. 144:3: "Lord what is man, that thou takest knowledge of him! or the son of man that thou makest account of him." In this Scripture we learn that God is accountable for man, but he's not accountable to man. On the contrary man is accountable to God. Aren't you glad you serve a concerned God who can account for you? God even has His eyes on the birds.

Brother's Keeper

In the Book of Philemon, we can clearly see that the Apostle

Paul was indeed his brother's keeper. On behalf of Onesimus, a slave of Philemon, Paul wrote a letter to the slavemaster urging him to receive Onesimus, not now as a servant but as one above a servant, a brother specially loved by the Apostle Paul. Paul explained to the slavemaster that Onesimus was his brother in the flesh and in the Lord. The Apostle said, "If thou count me therefore a partner, receive him as myself. If he hath wronged thee, or owest thee aught, put that on my account" (Phil. 1:17–18). Paul beautifully accounts for his spirituality here by standing up for Onesimus. It could have cost his life or friendship. But that was obviously irrelevant to Paul because he knew to "live is Christ and to die is gain." Let us stop being just nominal Christians. Let us become spiritual Christians and the world will see beauty in us.

Rulers Account for His Sheep

The Lord teaches us in Heb. 13:17 to submit to the authority of those who have the rulership over us because they watch for our souls. We should obey those who rule over us that they might give their account of us with joy and not with grief. Those who rule over fellow obedient, spirit-filled Christians will present joyful accounts of them to the Lord.

Accountability of a Certain Rich Man's Steward

Luke 16:1–8 teaches us that a certain rich man had a steward who had been accused of poorly managing his goods. The steward was accused of wasting his goods. So the rich man called in the steward and said, "How is it that I hear this of thee? Give an account of thy stewardship; for thou mayest be no longer steward" (Luke 16:2). "Then the steward said within himself, what shall I do? For my Lord taketh away from me the stewardship. I cannot dig; to beg I am ashamed. I am resolved what to do, that when I am put out of stewardship they may receive me into their houses. So he called everyone unto him and said unto the first, How much owest thou unto my Lord? And he said, a hundred measures of oil. And he said unto him, take thy bill and sit down quickly and write fifty. Then said he to another, and how much owest thou? And he said, a hundred measures of wheat. And he said unto him, take thy bill and write fourscore. And the Lord commended the unjust steward, because he had done wisely: For the children of this world are in their generation wiser than the children of light" (Luke 16:3–8).

The steward was obviously an unjust man. The word of God declares that if you make friends of the mammon of unrighteousness, when you fail, they will receive you into everlasting habitations. In conclusion the Lord also teaches us, "He that is faithful in that which is least is faithful also in much: and he that is unjust in the least is unjust in much" (Luke 16:10). If you have not been faithful to unrighteousness mammon, who will commit to your trust the true riches? The answer, my friend, is written in the wind.

Social Grace: Etiquette

The Lord Jesus said, "Behold I stand at the door and knock. If any man hears my voice . and opens the door for me I will come into him and eat with him and he with me." (Rev. 3:20)

XI
Social Grace: Etiquette

Being one of God's watchmen on the wall, I am observing globally some leaders in the Body of Christ, and I see a useful missing link that needs to be incorporated into their social behavior if they intend to be Christlike examples before the sheep whom they feed daily and before the observing world. It's the missing link of etiquette, and if a leader fails to acquire social grace from God's Holy Spirit, it will always be missing. Only God can give one social grace because the Bible says "All good gifts come from God, the Father of Light. There is no darkness in Him." Proper social behavior is an ordinance of God. In 1 Cor. 14:40 the Lord tells us, "Let all things be done decently and in order." I see etiquette in this Scripture. In Rev. 3:20 Jesus says, "Behold, I stand at the door and knock. If any man hears my voice and opens the door for me I will come into him and eat with him and he with me." I see etiquette in this Scripture also. Though Jesus is Lord, He's also a gentleman. He will not force His way into any man's secret chambers. In order for Jesus to enter a man's heart, He must be invited there as Lord of his life.

In this chapter, we're going to talk about the following items of utmost importance: etiquette at home and its usage in your ministry, application of etiquette everywhere and in everything, personal magnetism: its strength, the beauty of etiquette, the essence of etiquette, and the necessity of etiquette in Christian leadership. No one is exempted!

Etiquette at Home and Its Usage in Your Ministry

Every child should be taught etiquette from childhood through his early adolescent stage. If carefully taught he will never part from the teachings because its root system began during those formative years. It is therefore, the responsibility of the spiritual leadership of the home to teach their children etiquette by example. The spiritual leader of the home should not laugh at his daughter or son when he or she commits rude or discourteous acts. On the contrary he should correct them by any logical means necessary. The Holy Spirit of God will be his guide. The spirit will tell the parent which dose of medicine to administer to the child in each situation. Sometimes he will say use psychology. Sit down and talk only with the child. Other times He might say administer a good dose of castor oil. In other words, "don't spare the rod."

Let us not confuse using the rod with child abuse. When excessive use of force is used on a child, that's child abuse and it is also an abomination against God. But when the rod is applied intelligently with intent to teach the child a lesson he doesn't seem to grasp through dialogue and other forms of communication, it is called loving the child. God says, do not spare the rod, and I chasten all whom I love. Who among us has been God's counselor? As a result of parents sparing the rod, we now see much anarchy in the home, rudeness, disrespect for parents and the elderly. In some homes the children rule. They do whatever they feel like doing and the parents allow it. Many parents seem to find it easy to condemn or criticize other people's children but don't seem to be able to find room in their hearts to correct their own. The New Testament teaches us to take the log out of our own eyes first before we take the straw out of our brother's eyes.

Children should be taught to sit, walk, stand, eat, and to dress decently and mindfully. This is all a reflection of what they are inside. Tell the children that their overt behavior reflects their inner spirit. This is all *spiritual etiquette* and I dare you to confuse it with legalism. Children should be taught the value of manners in public and in privacy. If parents fail to teach them, they will surely become ill-mannered children. Unfortunately, there are many in our society today. Let the leaders also be an example of spiritual etiquette among the followers in his ministry. Many people imitate their leader even if the leader is wrong. This is ignorance! We must follow Christ, not earthly man. The

Apostle Paul said, "Follow me as I follow Christ." In other words one follows the *Christ* within his leader.

God himself orchestrated etiquette. Etiquette teaches us *how* it's done, *when* it is done, *where* it is done. We learn throughout etiquette the depth, breadth, and height in the spiritual realm. The Lord himself has prepared the table for us. Therefore, let us sit and dine with Him that we might prove what is that good and acceptable and perfect will of God.

Application of Etiquette, Everywhere, and in Everything

We need to practice etiquette at the dinner table, in conversation, on waiting lines, in crowds, etc. Why? We practice etiquette at the dinner table because it has a positive effect on our digestive system. How the food is served and our table manners can turn one on to the food or turn one against it. Don't be deceived, my friends, there are only two ways to do a thing. There is the right way and the wrong way. Now the right way may be manifested in many forms, but the results are positive. On the other hand, the wrong way may also be manifested in many forms and the results will always be negative. We should practice etiquette in conversation because failure to do so will cause confusion and misquotes.

In other words when one is speaking, the others should keep silent. It is impossible for everyone to speak at the same time. The Lord tells us in 1 Cor. 14:29–33 that the prophets should speak one by one. Speaking one by one is godly order. In Eccles. 3:7 God tells us that there is a time to keep silent and a time to speak. Waiting one's turn to speak should be a common practice, but it is not. It should only take common sense to do this. There is rudeness in conversation among Christians as well as among the people of this world. Waiting your turn to speak is a civilized thing to do if you're engaged in a dialogue. While you're speaking you should also be considerate of others who desire to speak.

And "let your speech be always with grace seasoned with salt that ye may know how to answer every man."

When one interrupts another's thought in order to dominate the conversation, nine times out of ten what he has to say will not address what the other conversationalist had in mind. Many leaders do not display a sense of etiquette when in dialogue with their followers. They're very rude to them and become very conclusive before they have listened to and analyzed what the follower has to say. Many godly followers will not attempt to tell the ministers about their rudeness because they know it will be in vain. Instead they take it all to Jesus and begin to look for another shepherd who is sensitive to all human beings and to whom they can easily relate.

We practice etiquette in waiting lines because the Lord has taught us to have patience. In public places we do not cut lines or get in front of someone because we're in a hurry. It can get you in trouble and might even cost your life in a rough city. Let us learn patience—wait our turn. While in crowds let us learn to be considerate of others and cooperative. You're all gathered there for a common cause. Why not cooperate and make it a lovely day?

Personal Magnetism of Etiquette

One who has social grace has the power to draw men nigh. Men begin to inquire. They want to know more about you. And it gives the leader an opportunity to tell them all about Jesus. It's the anointing of the Holy Spirit that breaks the yoke. Leaders who possess personal magnetism say to the inquirers, "In Him we live and move and have our being" (Acts 17:28). Personal magnetism is a gift from God to be used only for His glory and honor and not for the lust of the flesh.

The Beauty of Etiquette

God has created many beauties. However, one of the most

important social beauties He has created is etiquette. It makes living together, dining together, and socializing with each other a very pleasant experience. King Solomon said, "Pleasant words are like a honeycomb sweet to the soul and health to the bones." Many people appreciate pleasant folks. Some folk see godliness and pleasantness as a sign of weakness and will quickly exploit you if you allow it. But the good Lord teaches us to rejoice with those who rejoice with us. As King David said, "The joy of the Lord is my strength."

The Essence of Etiquette

The essence of etiquette signifies beauty, joy, social and spiritual correctness, decency and pleasantness. Without these factors no man could sincerely tell another to have a pleasant day. And at this juncture in my book, let me quickly say to each reader, "Have a pleasant day or evening. Carry social grace and etiquette with you wherever you go."

The Necessity of Etiquette in Christian Leadership: No One Is Exempted

Since God has commanded us to let everything be done decently and in order, we can say social grace and etiquette are two necessary tools we must use in order to accomplish decency and order. Without them both, we would fail God.

In conclusion of this chapter, I pray that every reader who lacks social grace will request it of God. And I pray that those who have it will use it so that God's name might be glorified even in their social circles. And let those who lack social grace in conversations say unto God as King David said, "Set a watch, O Lord, before my mouth; keep the door of my lips" (Ps. 141:3). Etiquette consists of conventional requirements as to social behavior. It is the properties of conduct as established in any class or community or for any occasion. There is a tremendous decline of etiquette in our society even among a good number of Christians. Unless Christians sharpen their tools and hearken to the voice of our God, they will become involved in this world system, which includes disrespect for authority, lack of etiquette, ethics, calling that which is good evil, and evil good. Let us repent globally and see the glory and mercy of the Lord.

Ethics in Christian Leadership

Rule
Of
Thumb:

"Do unto others as you would have them do unto you" (Mt. 7:12)

"Let not then your good be evil spoken of." (Romans 14:16)

"Abstain from all appearance of evil." (1 Thes. 4:22)

XII
Ethics in Christian Leadership

The human body that suffers from malnutrition will surely collapse if the dweller therein fails to obtain the proper medical treatment and nourishment. We may also apply the collapse theory to lack of ethics in Christian leadership. If Christian leadership lacks ethics, it will surely collapse also. Leadership astronomy centers around the Christlike ethical behavior of its leaders. Ethics in leadership is of utmost importance because "ethics" is a comprehensive word, which embraces the moral code of conduct of Christian leaders and their followers. Ethics is the stethoscope by which we test the blood pressure of Jesus flowing through our veins. The Holy Spirit of God has explicitly revealed to me that the blood pressure tests results are as follows: (1) Some leaders have spiritual high blood pressure (unethical behavior), (2) some leaders have spiritually low blood pressure (unethical behavior), (3) some leaders have spiritually normal blood pressure (ethical behavior).

God's Holy Spirit is our gauge and Jesus is the balm in Gilead, that great physician. He alone can cause your spiritually abnormal blood pressure to become normal. Let us strive to achieve excellence in ethics. The Lord says in His Word, "I can show you a more excellent way." Let the Lord show you that more excellent way. He wants to teach you ethics, but He will not do so unless you come to Him as a little child in your inquiry. One thing we must always remember is that Christian ethics is based upon divine truth. God has given every man a conscience that he might be able to discern right from wrong. When man yields his conscience to God, the Holy Spirit will help him to be the ethical servant God wants him to be. Now let's take a trip down ethics lane and hear what thus saith the Lord.

To the Ethical Leader

Just as there is ethics in secular fields, there is ethics in Christian ministry. When ethics is sincerely practiced, it can win the heart and respect of man. It is a powerful tool that draws people to your ministry or that causes men to respect your business or profession. The ministry that is founded upon moral principles or ethics is built upon Christ, the solid rock, and it will stand forever. It will endure testing. It will endure the winds and rain and stormy weather. Ethics is a system of moral principles. It's the rules of conduct recognized in respect to a particular class of human actions. We may also say that ethics includes the moral principles of an individual. Therefore, we should never forget Rom. 14:16: "Let not then your good be evil spoken of;" and how about 1 Thess. 4:22: "Abstain from all appearance of evil." These Scriptures are the epitome of Christian ethics. They are also the heartbeat of ethics. When your ethics fail to incorporate godly spirituality, it has no heartbeat at all. It ain't alive. It's dead, dead, dead!

Rule of Thumb

The New Testament teaches us in Matt. 7:12: "Do unto others as you would have them do unto you." What a wonderful way to live! If only the people of God lived by this Scripture, we could have a beautiful social and spiritual world. Unfortunately, many don't. Many people do unto others as they would ***not*** have them do unto them. When that which they do unto others is done unto them, they scream the loudest. They have no tolerance whatsoever. Let us pray that those to whom this is applicable will soon overcome. Make no mistake about it. Matt. 7:12 is a powerful word to this culturally diverse world. If the people of every culture listened to God's voice, there would be no racism because they would understand that the Almighty God created only one race and that's the human race. Man, through his vain imagination and with the guidance of Satan, who is the prince and author of all evil, created races and racism. God is no respecter of persons. Man looks at the outward appearance, but God sees the heart. Just as the Lord has given us technology, physics, geometry, and science to enable us to become a more industrialized society, He has also given us His Word to heal our social ills. Man's problem: He wants God's technology but doesn't want to obey His Word. In other words man wants to do his own thing. But God wants to do a new thing in man. God wants man to put his trust in Him, not in the technology he has allowed him to acquire. "Some trust in chariots and some in horses, but we will remember the name of the Lord our God" (Ps. 20:7).

Balanced Attitude and Aptitude in Ethics

The New Testament teaches us to "grow in grace and in the knowledge of our Lord and Savior Jesus Christ." In so doing the servant becomes a well-balanced spiritual person. Proper attitude and good conduct must precede growth in knowledge. If not, the acquiring of knowledge will puff up man. Man's hunger and thirst after knowledge must be to honor and glorify the omniscient God. Therefore, leaders must not allow men to work in their ministries only because they have a D.Th., Ph.D., or D.Min. degree. They must first be proven or tried to see if Calvary is behind those degrees. A D.Min. degree does not necessarily mean one is a spiritual person. However, application of divine truths one acquired on the road to his degree does indicate spirituality. When looking for a job, the attitude of the seeker gets him in the door, but aptitude ***keeps*** him there. Therefore, the formula for holding a job is attitude and aptitude = a job. One can learn a skill provided he's motivated and has the right attitude about learning.

Ethics in Christian Ministry

Christian ethics through the inspiration of the Holy Spirit

is the driving force that draws men to Christianity. We can easily understand why the Lord Jesus said, "A new commandment I give unto you, that you love one another. This do and they will know you are my disciples by your love." Loves says it all. Love is courteous; love is obedient. It submits to those who have authority over them; love is attentive; love follows instruction; love does not steal another minister's sheep; love does not envy. I hear it said and I agree completely: "Christianity is the religion of God's people and it brings us all to a state of common sense."

The Boundary Line

There is order in the house of God. Therefore, the spirit-filled organized leader must draw a boundary line between himself, his staff, and his followers. This is the chain of command. Though boundary lines are drawn, the leader must never isolate himself from his followers. He must be accessible to them in time of need. Boundary lines are drawn for ethical reasons and to enhance job production. There has to be a chain of command in ministry. You cannot let everybody do as he or she pleases. That's disorder. Everyone must be taught the chains of command and the consequences if the chains are broken. Breaking the chains is crossing the boundary line. Let us not confuse drawing a boundary line with dictatorship because it is not. On the contrary it is godly order and leadership.

Working Your Staff

Give your staff a job description and let not this job description be rhetoric alone. Train them if need be and let them work. If not, they will become bored. Sitting idly by while there is work to be done can bore anybody. If the staff member or follower is an elder, let him do an elder's job. He's not there only to occupy space. He accepted the post believing he would do the work of an elder—not sit there for the most part as a zero in the church.

When you betray a follower's confidence, you'll have some trouble trying to restore it. You must make every staff member feel important and treat them like he or she is somebody. And every believer is somebody in Christ Jesus.

If the follower is director of the usher board, let him direct it. If he is head of your evangelism ministry, let him direct it. Give your followers directions, not dictatorship. When they need you, be there for them! If you fail them, they will eventually leave you with a bleeding heart, knowing that you have not been completely just and straightforward with them. If you think men who work under you are fools, you're deceiving yourself. Please don't confuse meekness with weakness. For meekness is not weakness. It is strength under control. Leaders, God is watching you and you're accountable to God for your criminality (sin). God calls it sin. That's right. Sin against the Most High God is crime in the spiritual realm.

Fairness

Treat all your followers and colaborers justly because God is just. Make no exception. When you mistreat a follower, he remembers it, and the bond between you gets weakened if you as leader are the dictatorial type (one with whom one can not reason). Throughout the years I have observed much unfairness in churches and the society at large. I have seen so much of it, and it has really turned me against unfairness. I ask the Lord to keep me from the sin of unfairness and rudeness and give me the proper course of action to take whenever I am caught in the middle. We are the temple of the Holy Spirit.

The Ugly Face of Discrimination in Ministry

There is discrimination in ministry, whether men confess it or not. Some of it is racial; some of it relates to educational status; some of it is class; and some of it is economical. None of these giants have a place in the kingdom of God. The Lord looks at the heart of man. He resists the proud but exalts the humble. Discrimination, whether done consciously or unconsciously, is an atrocity. It's a great evil, and if discriminators fail to repent of their sins, the wrath of God will surely fall upon them. The Bible says the wrath of God falls upon the children of disobedience.

Sensitivity

I believe church leaders, just like secular leaders, should undergo some sort of sensitivity training. It would be ideal for those who work in a multicultural ministry. But it should not be limited to multicultural laborers, as I said previously. Many people in ministry roles from the top to the bottom do very rude things and don't think anything of it. There is no conviction whatsoever. Instead there is business as usual. If you confront them even in a godly way about the sins they have committed, they will get angry with you as if you are the one who sinned against them. They add insult to injury! What great wickedness! And what insanity! Some make appointments with you and fail to keep them and fail to apologize—even if it took you three hours to get there, and even if they could have called to cancel. Anyone with common sense should know that failures like that without apologies put a strain on the relationship and it changes the chemistry between you. Sensitivity is an excellent commodity to have. Lack of it can get you into a world of trouble. As leaders let us become more considerate of others and more patient. For the Lord is doing major heart surgery, and there is not going to be a bypass.

Tabernacle of a Christian Leader

THE BODY

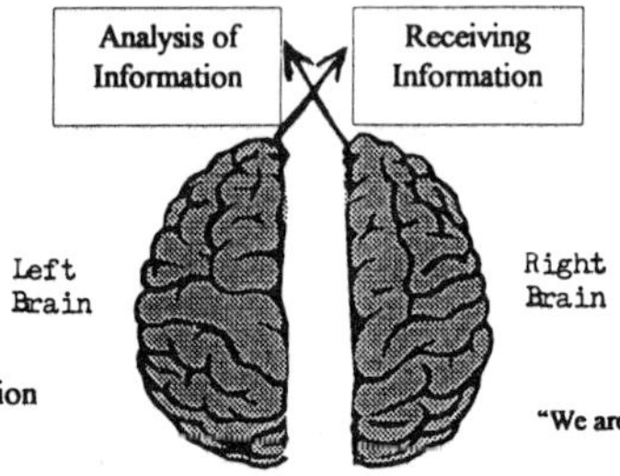

Sources of Information

1. Sight
2. Sound
3. Touch
4. Smell
5. Taste

"We are fearfully and wonderfully made." Ps. 139:14

Therefore we can never experience a feeling until we first have a thought.

THE SOUL

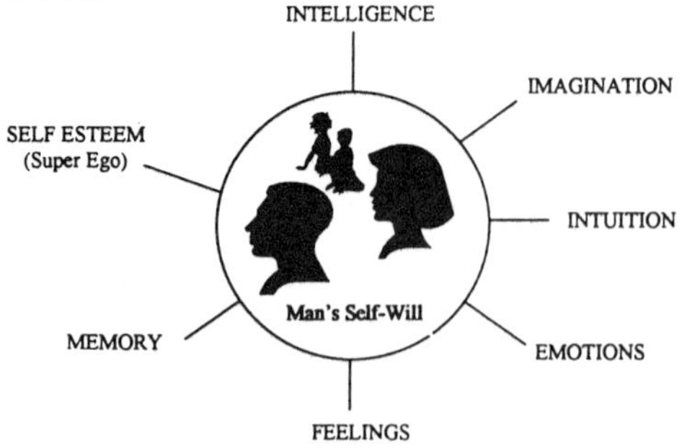

Know ye not that your body is the temple of the Holy Ghost which is in you, which ye have of God, and ye are not your own? (1 Cor. 6:19)

XIII

The Tabernacle of a Christian Leader

When scripturally learned man thinks of a tabernacle, he customarily thinks of the following three things: (1) the portable tabernacle of Moses, (2) the tabernacle of David, and (3) a place or house of worship. However, in this book, we are not going to focus our attention on any of these three tabernacles. Instead we shall focus our eyes on a fourth tabernacle called the tabernacle of a Christian leader. Now to the carnal mind, the fourth tabernacle is invisible, but to the spiritual mind it is quite visible! Only the spiritual minded can understand spiritual things. Let us first take a panoramic look at the tabernacle of a Christian leader and then a specific look at aspects that we might study their component parts.

Carefully examine the diagram at the beginning of this chapter, and you will become well acquainted with the tabernacle of a Christian leader. Your spiritual eyes shall behold the wonders of God's dexterity and love. The Old Testament teaches us that we are fearfully and wonderfully made. Furthermore the Most High God clearly states in 1 Cor. 6:19 "Know ye not that your body is the temple of the Holy Spirit which is in you, which ye have of God, and ye are not your own?" The Lord undeniably designed man to be a tabernacle of praise and worship unto Him.

Now let's take a look at the tabernacle of a Christian leader and see how it operates.

The Left Brain

Both the left and right brain take in information whether it be good or bad.

The Right Brain of the Tabernacle

The left brain analyzes the information taken in by the right. God is holy and just. Therefore, he has given man a conscience and an innate ability to know what is negative and what is positive, what is right and what is wrong. The Apostle Paul said in Acts 24:16: "Herein do I exercise myself to have always a conscience void of offense toward God and toward men."

Source of Information

Sources through which information enters the right and left hemispheres of the tabernacle are the five senses of the dweller: seeing, hearing, touching, smelling, and tasting. Though the creator has strategically given man each sense to use for his glory, failure to comply with God's wishes causes sin to triumph. Let's look at the sense of sight or seeing.

Sense of Seeing

An enormous amount of information taken in by the brain arrives there through the sense of sight. Through voluntary reading and through voluntary and involuntary observation, we obviously acquire sense of sight information. The acquired information may be useful or destructive. Nevertheless, the Word of God says, "It is not what goes into the mind; it is what comes out." In other words, it is what you do with the information that goes into your mind. Now some things a man observes or reads about can easily cause him to sin if he's not careful. Some writings or observations can lead another to repentance. For example: Usually, when a man commits the sin of lust, jealousy, or racism, it is through the sense of sight. Not until he mortifies the members of his body and brings his sense of sight before God's throne of Grace will he be able to use it for the purposes for which it was created.

Sense of Hearing

In the left brain, we also obtain an abundant amount of information through the sense of hearing. As intelligent human beings we should listen to and accept that which is logical and reject the illogical. What a man hears can be quite enticing. The information can have an adverse affect on him or it might have a positive one. In other words what a man hears can cause him to sin or cause him to live a godly life. This is why the Most High God says in the New Testament, "Faith cometh by hearing and hearing by the Word of God" (Rom. 10:17).

Touching

The sense of touching is very informational. Generally, through touching one can discern whether cloth is silk, wool, cotton or rayon; whether a ball is hard or soft; whether a thing is pleasant or unpleasant. The Word of God specifically commands us to "Touch not, taste not, and handle not the unclean things." Therefore, if we adhere to the touch-not law, we will avert the pitfalls of improper touching. Let's touch that which is clean and pure and let us touch not the unclean things, which spell sin.

Smelling

Through the sense of smell, we can determine whether a thing is pleasant or unpleasant. That is good and very useful for us as human beings. However, we must note that the sense of smell can also be used as a tool of Satan. Let me quickly cite one example: When fragrances are used to entice or lure someone into sin, the sense of smell is manipulated for evil purposes. But if the usage of the fragrance be not to cause one to sin, then and only then is it being used for the purposes for which God created it. Fragrances should never be used in attempting seduction. Instead they should be used only to give the body a pleasant smell. Remember, fragrances belong to God and we shall be held accountable before God, our Judge. The New Testament teaches "all things were made by him and without him was not made anything that was made."

The Sense of Taste

Everything that tastes good is not good for one's health nor spirit; everything that tastes bad is not bad for one's health. True it is that we use our sense of taste to determine if an item is hot, warm, or cold, sweet or bitter, salty or unsalted. But we need wisdom from God and the ability to reason that we might discern whether or not we should eat or drink an item. Let me give you some examples. Castor oil tastes bad. It tastes awful and the flesh tells us not to take it. But the spirit tells us to take it because of its medicinal value. It cleanses our system. So we tolerate the taste to obtain a cleansing blessing.

How about drugs and alcohol? To some people these products taste good, and as a result many folks have become addicted to the enemies of their souls and physical bodies. Everybody knows that an excessive consumption of alcohol can cause cirrhosis of the liver. Drugs and alcohol can also destroy the immune system and do tremendous damage to one's behavior. When a person is under the influence of these drugs, he becomes weak, and man's sinful weakness is Satan's opportunity. Satan takes control and man becomes a tool in the hands of Satan who walks around like a roaring lion seeking whom he may devour. How about pleasant-tasting products? Do we eat a product because of taste alone? No, we eat it because of its nutritional value and safety. However, if we consume an excessive amount of that which is nutritious, it can also have an adverse affect on us.

We've been talking about the left brain, the right brain, and the sources of information of the tabernacle of man. Now let's talk about the spirit of the tabernacle because man's spirit is the ruling force. God says, "He that rules his spirit is better than he that conquers a city." Though the right brain of the tabernacle takes in information holistically and the left brain analyzes it, it is the spirit of the tabernacle that makes the decision and acts upon it. The spirit symbolizes man's will. That's why the Lord Jesus taught us to pray "not my will but yours be done." To please God we must commit our will unto him. We have a free will and He will not force us to commit our will to him. Be that as it may, the will of the tabernacle is the commander-in-chief of all military zones. "For God did not give us the spirit of fear, but of power and of love and a sound mind" (2 Tim. 1:7). Unless there is action, there can be no reaction. The spirit reacts to the information that the left brain of the tabernacle analyzes.

Now let us take a look at the soul of man of the tabernacle. Satan, the adversary of God and man, is struggling to conquer your soul. You need to become well acquainted with the avenues on which he walks and works.

Intellect

When God created man, he gave him an intellect, a capacity to learn or acquire knowledge. If a man fails to stimulate his intellect through learning exercises, he will pay the consequences. The consequential price can be quite costly. Let us examine what God has to say about knowledge. In 1 Cor. 8:1, he says, "Knowledge puffeth up, but charity edifieth." In Hos. 4:6 the Lord says, "My people are destroyed for lack of knowledge." In these two Scriptures, we see the negative effects related to knowledge. It has the power to puff up or make one proud; one who lacks knowledge can easily be annihilated or destroyed. What is the solution then? The Lord gives us the solution in 2 Pet. 3:18: "But grow in grace and in the knowledge of our Lord and Savior, Jesus Christ." Unless a man acquires wisdom, his knowledge will not have a positive effect. Wisdom is the application of all knowledge.

Imagination

God states in Prov. 23:7: "For as he thinketh in his heart so is he" and in Prov. 4:23: "Keep thy heart with all diligence; for out of it are the issues of life." As we can clearly see here, man's creator has given him the ability to think and form ideas. From man's ideas and decisions come all issues of life. No issue can take place before an idea and decision. In psychology we call this the law of cause and effect. Therefore, we must feed our spirit the nutritious word of the Most High God, which enables us to form logical ideas and make godly decisions.

Intuition

According to the *Random House Dictionary of the English Language,* intuition is "a keen or quick insight; the quality or ability of having such direct perception or insight; pure untaught, noninferential knowledge." Since intuition deals with keen insight and untaught noninferential knowledge, it is obvious that a man's intuition was created within him by God. By adhering to our godly intuition, we can avert many pitfalls of life. When God created man, he gave him an intellect and intuition. Now as I stated previously, the intellect deals with the acquired taught knowledge, but the intuition deals with keen insight and one's untaught knowledge, which comes directly from God, the Father of Light.

Emotions

A person's response to life situations is called emotion. Sometimes it's joy; sometimes it's sorrow; sometimes it's hatred; sometimes it's fear; sometimes it's love. Be that as it may, they're all emotions. Though the Lord has given man the ability to respond to situations based on Gal. 5:23, He has also given him a fruit of the spirit called temperance, which means self-control. This fruit enables a man to have emotional stability. Emotions and temperance = emotional stability. Lack of temperance will surely lure one into sin. Though one's emotions are always *self-directed,* they take wings and fly to others in our environment. In other words our emotions are capable of having positive or negative effects on others. In Job 3:1–26, we see how emotions affect self; in Ps. 137:1–6 we see clearly the effects of emotions on a nation; in Gen. 49:1–28 we see the effects of emotions on a family; in 1 Sam. 25:24–25 we see the effects of emotions on a mate; and in Ruth 1:16–18, we see the effects of emotions on foreigners.

Feelings

Feelings are sensibilities and when one's feelings are touched, there is always an emotional response. Feelings are always directed toward people, places, and things. Therefore, it is of utmost importance to bring our feelings to the throne of Grace that we may obtain mercy and find grace to help us in time of need.

Memory

God is an omniscient God. He knows everything. In other words He's all knowledgeable. He remembers everything and even knows it before it takes place. Though man doesn't remember everything, God did give him in creation the ability to recall or remember that which he learned. Memory is a very useful tool of the soul. Without a memory capacity, a man would be incapable of learning that which he is taught. Moreover, there would be no use for teachers. Our God is a God of wisdom and man is a unique creature. He is a very special creation of God.

Self-Esteem (Superego)

Self-esteem is an objective respect for or favorable impression of oneself. It's a spiritual value of oneself. The New Testament teaches us Christ within us is the hope of glory. Additionally, King David said, "For I am fearfully and wonderfully made: Marvelous are thy works; and that my soul knoweth right well" (Ps. 139:14). God also annexed superego, which is connected to self-esteem, to our soul. Superego is part of the psychic apparatus. The ego mediates between it and the id drives. The superego contains social ideals acting as a conscience, which may be partly conscious and partly unconscious. We thank God for giving us a conscience and self-esteem. God, the righteous judge, will judge all men by their conscience. For men will be judged according to the amount of light that they have received (John 8:7–9).

In conclusion of this chapter on the tabernacle of a Christian leader, the Holy Spirit wants to make it very clear to the readers that the physical bodies in which we dwell are merely tenements for the invisible soul and spirit. The New Testament teaches us that it does not yet appear what we shall be like. For we shall be like Him. Before we enter the portals of glory, we shall be given a celestial body. Let us rejoice. And let us remember that our bodies are temples of the Holy Ghost and we shall individually and collectively become the tabernacle of praise that the Father of light desires. God has called us to holiness.

Power Tools of Dynamic Leadership

Love

Joy

Peace

Gentleness

Goodness

Faith

Meekness

Temperance

GOD'S WORD
HEB. 4:12

Baptism of the Holy Spirit
ACTS. 1:8

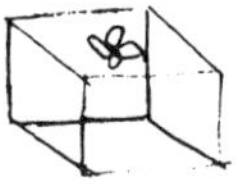

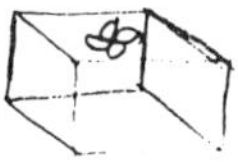

GIFTS OF THE SPIRIT

Revelation Gifts	Utterance Gifts	Power Gifts
Word of Knowledge	Tongues	Faith
Word of Wisdom	Interpretation of Tongues	Healings
Discerning of Spirits	Prophecy	Working of Miracles

XIV
Power Tools of Dynamic Leadership

Without power from on high, the frail human being could do absolutely nothing. For the Lord remembers man's frame, which He molded from the dust of the earth. God remembers that men are frail and made of clay. Therefore, they need empowerment from their creator. The Holy Spirit of God is men's power base. He's the source of power. Let's take a good look at Acts 1:8. It is written: "But ye shall receive power, after that the Holy Ghost is come upon you; and ye shall be witnesses unto me both in Jerusalem, and in all Judea, and in Samaria, and unto the uttermost part of the earth." In this Scripture we can clearly see the effects of the Holy Ghost when He comes upon the disciples of the Lord. They receive power and witness for Him unto the uttermost part of the earth. In Him we live and move and have our being (Acts 17:28).

In this chapter we shall study the power tools of dynamic Christian leadership. The Holy Spirit of God shall unveil the tools and the dynamics of their usages. We are also going to discuss their importance and see how these tools are reflected in a leader's or believer's supposedly Christlike model lifestyle.

Now let's take a close look at the basic power tools of dynamic Christian leadership.

God's Word, a Tool of Power

God is the source of the Bible and all Scripture is given by His divine inspiration. "Inspiration" means that the Word was God-breathed or divinely created. The Lord purposed that it be beneficial to us as a governing or guiding factor in our lives. Second Tim. 3:16–17 says that "all Scripture is given by inspiration of God, and is profitable for doctrine, for reproof, for correction, for instruction in righteousness." Though God is the author of the Word, He used man as a vehicle to make known His revelation. King David as well as King Samuel, Jeremiah, Daniel, Ezekiel, and Isaiah (authors of the Old Testament Scripture) were all prompted to speak by the Holy Spirit (Acts 1:16 and Acts 28:25). God's Holy Spirit is the source of infallible revelation.

The Word of God is not devised by man's intellect, and it is not subject to private interpretation (2 Pet. 1:20). God's Word is a source of power and a very useful tool for the servants of the most High God. It's a dividing power, as a "sword" (Heb. 4:12). It's a reflecting power, as a "mirror" (James 1:22–25); it's a cleansing power as a "cleanser" (Eph. 5:26; John 15:3); it's a reproductive power, as a "seed" (1 Pet. 1:23); it's nourishing power, as "food" (milk of the word—1 Pet. 2:2); guiding power as a "lamp" (Ps. 119:105 and 2 Pet. 1:19) and the Word of God is the power of God unto salvation (Rom. 1:16).

The Fruits of the Spirit

When we receive the Holy Spirit, we receive the enabling power to bring forth fruit in due season (Gal. 5:22 and Ps. 1:3). The fruits of the spirit are love, joy, peace, long-suffering, gentleness, goodness, faith, meekness, and temperance; against such there is no law. Now the word "fruit" is used here metaphorically to mean "result" or "outcome." The results or outcome in a leader's or believer's life should be the nine virtues listed in Gal. 5:22. The New Testament teaches us that the Lord Jesus is the vine and we are the branches and apart from Him, we can do nothing. "Vine" denotes fruit as being the visible expression of power working inwardly and invisibly, the character of the fruit being evidence of the character of the power producing it. The fruit of the Spirit is extremely important because only by man's fruit will the world know that He's a servant of the Most High God.

Gifts of the Spirit

According to 1 Cor. 12, there are nine gifts of the Holy Spirit and they can be grouped into three major divisions:

I. Revelation gifts
  A. Word of wisdom
  B. Word of knowledge
  C. Discerning of spirits
II. Utterance gifts
  A. Tongues
  B. Interpretation of tongues
  C. Prophecy
III. Power gifts
  A. Faith
  B. Healings
  C. Working of miracles

Perhaps you might ask, "Why are the gifts given to the church?" Well, here is the answer. Those precious gifts of God's Holy Spirit are given to the church for the following reasons: A. For the common good (1 Cor. 12:7); B. Edification (1 Cor. 14:3,5); C. For convictions and convincing confirmation (1 Cor. 14:24–25); D. Instruction and exhortation (1 Cor. 14:31); E. Confirming and establishing ministry (Acts 9:15; 13:2, 1 Tim. 1:18). The Holy Spirit of God gives the gift to the leader according to the divine will of God. To some He gives one gift and to some He gives two gifts or more. If the gifts are used wisely, they multiply the power in the leader's ministry.

Usage of Power Tools

The Lord gives power tools to His servants for edification and perfecting of the saints. Without the gifts of these power tools, no man could lead anyone to Christ. Without the godly gift of power bestowed upon the leader, he could do nothing for the

Lord. The challenge to man is that he must decide whether he will use that God-given power to glorify God, the giver, or will he seek praises and the applause of men.

The Importance of Power Tools

One needs power to do every imaginable thing. For example: a man needs power to sit, walk, stand, think, to imagine, to reason, and to plan. With the above thought in mind, it is obvious that a man needs power to lead. What then is the importance of the power tools? The power tools enable the leader to discern, to make decisions, to lead the flock of God intelligently, to analyze, and to live a Christlike model lifestyle before those whom he is leading.

Some Reflections of Power Tools

If a servant of the Most High God has God-given power, it will be reflected in his character, and in his overt behavior. Some of the qualities that he will reflect are:

1. Spiritual strength (Josh. 1:6–7, 1:9)
2. Obedience to the Lord (I Sam. 15:22, Acts 5:29, Jer. 7:23, Col. 3:22)
3. Virtue, fear of the Lord, honesty, a dislike for greed (Exod. 18:21) and,
4. Meditation on the Word of God (Josh. 1:8).

Be that as it may, power tools in the leader's life must construct a reflecting mirror, because faith without works is dead. In other words his life must reflect what he believes (James 2:22–26).

In conclusion to this chapter, the author would like to stress that knowledge of God's Word, wisdom, fruits of the Holy Spirit, gifts of the spirit, understanding, the baptism and filling of the

Holy Ghost are the basic power tools of the leader. Without them he could not be victorious. Just as the carpenter needs tools to build a house, so does a leader need tools to construct his leadership.

Wait on the Lord

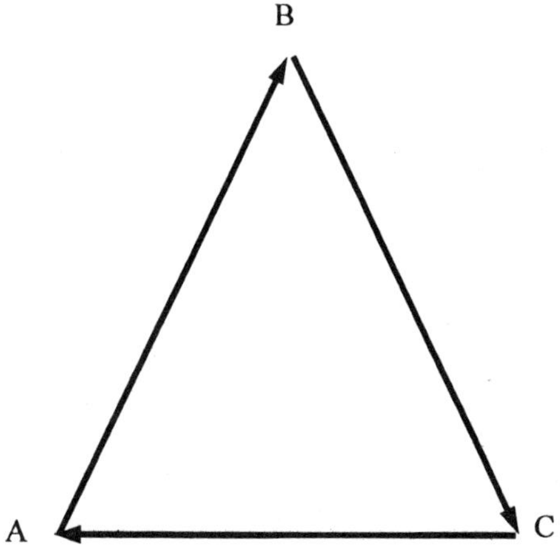

Rule of Thumb: Never attempt to move from point A to point B nor from point B to point C without consulting the Lord. Remember you're a servant in the Lord's vineyard.

"But they that wait upon the Lord shall renew their strength; they shall mount up with wings as eagles; they shall run and not be weary; and they shall walk, and not faint." *Is. 40:31*

XV
Wait on the Lord

Being a zealous laborer in God's vineyard demonstrates one's nobility. In fact working for the Lord is the most noble task in the whole wide world. However, when a zealous vineyard laborer tries to move from point A to point B or from point A to point C without seeking permission or wisdom from God, he causes spiritual friction in the vineyard. Godly laborers must learn to seek counsel from the Lord and wait on His reply. Moving along from one point to another without God's stamp of approval will surely result in failure. If you are trying to advance to point B from point A but can't because there are obstacles in your path, then you must wait on the Lord. You must pray and wait.

The New Testament teaches us that the Lord will go before us to make the crooked place straight. The battle belongs to God and He will give us the victory if we wait on Him. God says in Isa. 40:31, "But they that wait upon the Lord shall renew their strength; they shall mount up with wings as eagles; they shall run and not be weary; and they shall walk and not faint." The Most High God also teaches us that we should not be anxious for anything, but with prayer and thanksgiving, we should let our petitions be known.

The Holy Spirit has strategically included this chapter in the book because He knows that so many people find waiting on the Lord to be a difficult task. In the following paragraphs, I shall elaborate on (1) patience or long-suffering in relations to waiting; (2) faithfulness in relation to waiting; (3) tribulation in relation to waiting; and (4) God's glory in relation to waiting.

Patience or Long-Suffering

The righteous God has asked man to wait on Him, knowing that He has given him the power and the ability to do so. One of the fruits of the Spirit found in Gal. 5:22 is patience. This fruit enables man to wait on the Lord. Patience is developed as the servant waits, and the seeds of patience will bring forth fruit in due season.

Faithfulness

Based on Gal. 5:22, God also gives man the fruit of the spirit called faithfulness, which enables him to be diligent and true day after day and night after night. It gives man the capacity to serve God in season and out of season. Man has the ability to wait on the Lord if he wants to. However, his pride and eagerness to advance sometimes get in the way of waiting and also in the way of faithfulness.

Tribulation

I have news from the *Good News Holy Bible* for the unfaithful and the impatient. According to Rom. 5:3–4 tribulation worketh patience. Yes, this is God's medicine to cure one's lack of patience. The Word also tells us that patience worketh experience. And experience worketh hope.

God's Glory

What we are and all that we ever will be we owe to Calvary. Based on Col. 1:27, "Christ in you is the hope of Glory." The New Testament also teaches us that our light affliction, which is but for a moment, worketh for us a far more exceeding and eternal weight of glory (2 Cor. 4:17). Now it is true that the Lord takes some of us through the water and some through the fire. It is also true that He is nigh during the time of our great trials and tribulations. He takes us through the water to cleanse us; He takes us through the fire to purify us so that we might come forth as pure gold. Let's tune in to God in Isa. 43:2: "When thou passes through the waters I will be with thee; and through the rivers they shall not overflow thee; when thou walkest through the fire thou shalt not be burned; neither shall the flame kindle upon thee. For I am the Lord thy God, the Holy one of Israel. I have redeemed thee and I have called thee by thy name and thou art mine."

May the children of the vineyard wait on the Lord so that they might bring forth pleasant fruit in due season. Our God is a rewarder of those who diligently seek Him. He will bountifully recompense His faithful vineyard servants. So let us learn to labor and learn to wait. The vineyard belongs to God. We must learn to say like the prophet Job, "I shall wait until my change comes. I know my redeemer liveth. He lives within my soul." Are you hearing me?

"For I know that my redeemer liveth, and that he shall stand at the latter day upon the earth: and though after my skin worms destroy this body, yet in my flesh shall I see God: whom I shall see for myself, and mine eyes shall behold, and not another; though my reins be consumed within me" (Job 19:25–27).

Debate and Strife

"And Abraham said unto Lot,¨Let there be no strife, I pray thee, between me and thee and between my herdmen and thy herdmen; for we are bretheren.¨" Gen. 13:8

XVI

Debate and Strife

As I look over the mountains of leadership, I see much debate and strife among God's leaders and among world leaders. One does not need to possess the gift of Word of knowledge nor the gift of prophecy to be cognizant of this fact, for it can be clearly seen with the natural eyes. God opens up His windows of revelations in the supernatural realm only for the things we can not see in the natural. Over the mountains of leadership, I see competition among leaders, rudeness, coldheartedness, manipulation. However, I want to remind you that the Lord said in His word that "every mountain and every crooked place shall be made straight; every hill and every valley shall be brought low."

The Lord Jesus said, "Behold I come quickly and my reward is with me." He will also do a quick work in his priesthood prior to his return. God's end time priesthood shall be greater than the former because the Holy Bible tells us that the glory of this latter house shall be greater than the former (Hag. 2:9). God's end time priesthood will not be men of debate and strife but pleasant men of faith and inner strength. They will be men with a discerning of spirit. Praise and thanks be to God because it's beginning to rain now from the throne of the Father. Jesus is saying "Let him that thirst come to me and drink." And if we drink that distilled water that the Lord gives unto us, we will never thirst again, nor will we study war or debate and strife anymore.

Let me tell you a few surprising stories about debate and strife directly from the Word of God:

Debate Story of Who Will Be the Greatest Disciple

Though the disciples knew about the Lord Jesus' approaching sacrificial death, instead of being very supportive, they argued about which of them would be the greatest in the kingdom. What insensitivity! The Good Lord is preparing to give His life as a ransom for the human race and men of God debate over who will be the greatest among them. O God, O God, O God! Forgive us! If men had said to the Messiah, "From the depth of our hearts, we appreciate what you are about to do for us. And we pray that the Most High God will give us the grace to continue your works until thou returneth," it would have sounded pleasant to heaven's ears. This story is supported by chapter 22 of Luke and the key verse is Luke 22:24: "And there was also a strife among them, which of them should be accounted the greatest?" Who is the greatest in the kingdom? The Lord Jesus skillfully answers this question in Matt. 18:4: "Whosoever therefore shall humble himself as this little child, the same is greatest in the kingdom of Heaven."

The Sharp Contention between Paul and Barnabas

Let's tune into Acts 15:36–41 and we will get the whole story:

> And some days after Paul said unto Barnabas, let us go again and visit our brethren in every city where we have preached the Word of the Lord, and see how they do. And Barnabas determined to take with them John, whose surname was Mark. But Paul thought not good to take him with them, who departed from them from Pamphylia and went not with them to the work. And the contention was so sharp between them, that they departed asunder one from the other. And so Barnabas took Mark and sailed unto Cyprus; and Paul chose Silas, and departed, being recommended by the brethren unto the Grace of God. And went through Syria and Silicia, confirming the churches.

What a surprising contention! Paul's decision not to carry Mark on the missionary journey should have quickly settled the issue because Mark had deserted them on the previous mission. And he obviously had shown no signs of repentance. Yet Barnabas, a mature servant of the Lord, refused to see Paul's point of view. This is the mystery of iniquity. Unfortunately, men of great stature in the Body of Christ still have heated arguments over such things. To make matters worse, some do it in the presence of believers and nonbelievers. God has called his people to be a beacon of light in this world of darkness. Debate and strife won't draw men to God, but a Christlike character will. All Barnabas needed to do in that situation was to ask himself the rhetorical question: "What would Christ Jesus do in this situation?" Surely God would have come to his rescue. The Word of God says "Where there be two or three gathered in my name there I am in the midst of them."

Before concluding this chapter, the author wishes to make it crystal clear that we have debate and strife in the church and in the world because men violate the spiritual laws of God. Whenever we violate the laws of God, there will be friction. For example: If you fail to do unto others as you would have them do unto you, the abused will react defensively to your offensive behavior. The following are some common elements that cause strife: competition, manipulation, jealousy, rudeness, coldheartedness. The strife solution is the new commandment that the Lord Jesus gave us, found in John 13:34–35: "a new commandment I give unto you, that you love one another; as I have loved you, that ye also love one another. By this shall all men know that ye are my disciples if ye have love one to another."

SHEPHERDING

Psalm 23

XVII
Shepherding

When I was in the first grade down South, I learned a nursery rhyme at school titled "**Little Boy Blue**":

> Little Boy Blue, come and blow your horn,
> The sheep are in the meadows and the cows are in the corn;
> Where is the little boy who looks after the sheep?
> He's on the hay stack fast asleep.

What a revelation! The sheep are in the meadows and the cows are destroying the corn, while their shepherd lies soundly asleep on a haystack! Well, that ain't shepherding! That ain't shepherding at all! That's negligence!

Unfortunately, the attitude of a great number of God's earthly shepherds is very similar to that of Little Boy Blue. Instead of steering the sheep in the paths of righteousness, they are teaching and preaching a watered-down gospel. Many shepherds are conforming to this world's corrupt system in order to gain favor and to appeal to those who have "itching ears." Nevertheless, the Word of the Most High God teaches us to "watch and pray." In other words the shepherds must maintain a godly watch over their assigned sheep. They must love them and care for them. And they must pray continuously for them.

At this moment I want to explain to you the meaning of the word "shepherd." Afterwards we shall discuss the word "shepherding." Now "shepherd" is the root word. Therefore, we must begin there.

Shepherd

The Old and New Testaments use the word "shepherd." It means one who tends herds or flocks—one who cares for the sheep of the Lord. The Hebrew and Greek translations for the words "shepherd" and "pastor" are synonymous.

Shepherding

One cannot do effective shepherding labor for the Most High God until he gains a mastery of the meaning of the root word, "shepherd." No, I don't mean an academic mastery alone. For I am talking about a united spiritual and academic mastery. If one understands the meaning of "shepherding," is called by God to shepherd the sheep, obeys God's call, and is equipped to shepherd, then he can do a magnificent shepherding job. ***Shepherding*** is leading the sheep in God's paths of righteousness. It's teaching them and preaching them the undiluted Word of God. It's reassuring the sheep that the life, suffering, and death of Christ Jesus were all an atonement for the sins of fallen man. It led to the reconciliation of God and man. Shepherding is caring for the sheep.

The Invisible Chief Shepherd

We cannot see our chief shepherd with our natural eyes. But one day we shall behold Him! We shall see Him face to face. "And when the chief shepherd shall appear, ye shall receive a crown of glory that fadeth not away" (1 Pet. 5:4). The New Testament teaches us that the Lord Jesus is the good shepherd (John 10:11–14); Jesus is the great shepherd of the sheep (Heb. 13:20); Jesus is shepherd and bishop of our souls (1 Pet. 2:25). We refer to the Lord Jesus as the invisible **Chief Shepherd** because the earthly shepherds whom we shall discuss next receive their guidance from Him through the Holy Ghost.

Pastors, the Visible Under-Shepherds

Although the Holy Bible tells us that the Lord is our shepherd, God has also placed under-shepherds in the Body of Christ here below to minister to spiritual Israel and to lost souls. The Apostle Paul said, "My purpose is that of saving some." Based upon Eph. 4:11, pastors or shepherds were given to the church as a gift from above. In the Old Testament, the prophet Jeremiah prophesied that the Lord would set up shepherds over His people (Jer. 10:2). Therefore we must obey God's under-shepherds because they watch for our souls (Heb. 13:17). Now some of God's exhortations to the under-shepherds are:

1. Watch ye and stand fast (1 Cor. 16:13),
2. Continue in prayer and watch (Col. 4:2),
3. Let us not sleep, but let us watch (1 Thess. 5:6),
4. But watch thou in all things (2 Tim. 4:5),
5. Obey them that have the rule over you, for they watch for your souls (Heb. 13:17).

Pastors or shepherds are a tremendous blessing to the world! Without them the sheep would be consumed by savage wolves.

The Shepherd's Duties toward His Flock

Being a good shepherd is not an easy task. Nevertheless, the Almighty God enables His shepherds to do the work that He assigns them. Some duties or responsibilities of a shepherd are:

1. Feed the sheep knowledge and understanding (Jer. 3:15). God's word is true. He says in Jeremiah, "I will give you Pastors according to mine heart, which shall feed you with knowledge and understanding."
2. According to Eze. 34:2–4, a shepherd's duties are to heal those who are sick.

3. Bind up those who are broken.
4. Bring back those who were driven away.
5. Seek for those who are lost.
6. Not rule them with force or cruelty.
7. A shepherd should take oversight of his flock not by constraint but willingly; not for filthy lucre, but of a ready mind (1 Pet. 5:2). The shepherd does his work for the honor and glory of the Lord (Col. 3:23).

Description of a Good Shepherd

The Lord gives us a graphic description of a good shepherd in the New Testament. Let's tune in to what God says. In Gen. 31:38–40, God says a good shepherd is faithful; in 1 Sam. 17:34–36 a good shepherd is fearless; in Luke 15:3–6 we learn that a good shepherd is unselfish; in Genesis 33:13–14 we learn that a good shepherd is considerate; and in Luke 2:8–20 we learn that a good shepherd is believing. The shepherd believes in God and in the job that the Lord has assigned him. Now if we're going to do shepherding for the Lord, we need to learn all about the characteristics of a good shepherd. Knowledge about the characteristics of a bad shepherd can also serve as a warning to us. We can avert the pitfalls of bad shepherding if we desire.

Description of a Bad Shepherd

The Holy Ghost has included the description of a bad shepherd within this chapter to warn us. Bad shepherds will suffer tremendous consequences on God's Judgment Day. If any shepherd who reads this book is guilty of bad shepherding, I pray that you will repent today that you might escape the wrath of God. In Ezek. 34:1–10 the Lord says a bad shepherd is unfaithful; in John 10:12,13 God says a bad shepherd acts cowardly; in Isa. 56:11–12, God says a bad shepherd is selfish;

in Exod. 2:17,19 God says a bad shepherd is ruthless; and in Jer. 50:6 God says a bad shepherd is unbelieving.

Meet the Sheep of God

Based on Ps. 79:13 and Ps. 100:2, God's people are referred to as sheep. Reference to God's people as sheep can be found throughout the Holy Bible. The sheep of God are Christian believers who follow the Christlike model shepherds. The sheep also know the voice of their shepherd; they eat the food given them by the shepherd; they heed their shepherd's warnings, they bear fruit and they allow the shepherd to shear them. As you can clearly see, the sheep also have responsibilities. Can you picture a shepherd leading God's sheep without giving them any responsibilities? No, No, No!

Types of Sheep

A shepherd or pastor needs to know what types of sheep he can possibly encounter in the pasture. All sheep are not good sheep. Some are bad. Therefore, the shepherd needs to be well prepared and warned before he goes among them. Now I am going to introduce to some readers and present to others five types of sheep.

Solitary Sheep

Let's begin with the *Solitary Sheep*. Solitary is an excellent name for them because they stay away from the rest of the flock; they think the rest of the flock is too mature for them and they are always sick.

Hermit Sheep

They stay away from the rest of the flock because they don't want to be sheared; they don't want the pruning of the Holy Spirit; the wool grows over their eyes and they can't see to protect themselves; the wool weighs them down so that soon they can't rise and the wolves come in and devour them; they refuse to be clipped by the shepherd; they begin to influence others.

The Wandering Sheep

The Wandering Sheep are never content with things as they are. Whatever God is doing, it is never quick enough, never large enough, or good enough to please them. They always talk about the way it used to be. They have a tendency to bring discontentment to other sheep, and they are capable of producing other wandering sheep.

Judas Sheep

They lead others to the slaughter just like Judas Iscariot led our Lord Jesus to be slaughtered. Satan has blinded them and they are deceived by their own stubbornness. Their opinions are always contrary to the shepherd. They slander the shepherd and carry tales abroad. They are unruly, wild, and have no love for the flock.

The Contented Sheep

The contented sheep are the true sheep of God. You know the Apostle Paul said, "I have learned in whatsoever state I am, therewith to be content" (Phil. 4:11). The contented sheep are innocent, and they trust the shepherd. They are mild, tenderhearted, and gracious. They are very patient, industrious, useful. They are so very submissive to their shepherd. I pray that none of God's shepherds will grow weary when they meet the discontented solitary, hermit, wandering and Judas sheep. Let the shepherds here below look to the Hills of Zion for wisdom, and the chief shepherd will instruct them. The chief shepherd watches over his under-shepherds. Jesus, the bright and morning star, is the astronomy of Christian Leadership.

Wolves in Sheep's Clothing

God tells us in Matt. 7:15 that there are wolves in sheep's clothing; and in Matt. 7:16, God also tells us how to identify them. He says, "By their fruit ye shall know them." Which fruit? The fruit of their wolflike spirit. All you need to remember is the fruit of the Holy Spirit based on Gal. 5:22: "love, joy, peace, gentleness, goodness, meekness, temperance, faith, patience." A wolf in sheep's clothing cannot produce this fruit. The Holy Spirit manufactures this fruit only in true believers. Shepherds and the sheep should ask God for the gift of discerning of spirits so that Satan, the adversary of God and man, won't be able to deceive them through his wolves in sheep's clothing. Wolves in sheep clothing are earthly servants of the devil. However, God says in His word, "Greater is He that is within you than He that is within the world." God has strategically put the shepherd in the Body of Christ to prevent the dispersion of His sheep (Matt. 12:29 and Zech. 13:7). God wants his shepherds to comfort and bring the sheep light in moments of darkness.

When the Lord calls, just say YES!

"Many are called but few are chosen."
Mt. 20:16

XVIII

When the Lord Calls, Just Say "Yes"

According to the New Testament, the Lord Jesus taught, "As the Father has sent me even so send I you" (John 20:21). That means God has a mission for every born-again believer. He wants to equip and prepare you for your specific *assignment* or *assignments*. Just as the Lord Jesus had a mission, so do we. Therefore, when the Lord calls, just say "yes." Do not give Him excuses because He will not accept them. Perhaps you might think you *can't* do what the sovereign Lord is calling you to do, but He knows you *can*. In Phil. 4:13 God declares we can do all things through Christ which strengtheneth us. And in Ps. 139:14 God declares we're fearfully and wonderfully made. We're unique! God doesn't duplicate His creatures. Though we have failed God miserably, He designed us all to do a specific task. In our weakness the Lord makes His strength perfect. Stop hiding behind your excuses. When the Lord calls you, just say "yes."

Now let's take a *telescopic view* at the calls of the Lord to some of His chosen vessels of honor and see how the stars responded. Note: I call the selected vessels of honor stars because they are well known by Bible students throughout the world.

A View of Adam

Perhaps you never thought of Adam as a leader, but he was. He was chosen by the Most High God to lead humanity (Acts 17:26). Adam was God's first human creation. After his creation God put him in the Garden of Eden. Knowing that it's not good for man to be alone, the Lord gave him a wife, Eve. Though Adam was the leader, the Lord blessed them and told them to be fruitful and multiply and replenish the earth. "And let them have dominion over the fish of the sea, and over the fowl of the air and over every living thing that moveth upon the earth." God put Adam in that Garden and commanded him, saying "of every tree of the garden thou mayest freely eat; but of the tree of the knowledge of good and evil, thou shalt not eat of it; for in the day that thou eatest thereof thou shalt surely die."

Now let's see how Adam responded. He responded brilliantly to the call of the Lord and was very faithful. God brought every beast of the field, every fowl of the air unto Adam to see what he would call them. And Adam gave names to all of them. When God presented the female creation, Adam called her woman. He said, "This is now bone of my bones and flesh of my flesh; she shall be called woman because she was taken out of man." Seemingly, Adam loved God, loved his wife, and loved his call.

But one day the garden serpent (the devil) beguiled Eve; then Eve persuaded Adam to eat of the tree of the knowledge of good and evil. When Adam yielded to his wife instead of obeying the voice of God, that was indeed a turning point in human history. Because of the sin committed in the Garden of Eden, the parents of humanity fell from grace and their children became victims of circumstance. In order for man to become reconciled, God in His mercy and love sent the Messiah (Gen. 2 and 3). Adam enjoyed a paradise but through his weakness, paradise was lost. Nevertheless, through Jesus Christ, our Lord and Savior paradise will be restored some sweet day. Hallelujah! We believe in someday!

View of Noah

Though Noah lived in a very corrupt society, he was a man of righteousness. During his time, man's wickedness was so great until it grieved God's heart that He had ever made man. So God decided to destroy all flesh. However, Noah, a preacher of righteousness, found favor with God. The Lord told Noah to build an ark because He was going to bring a flood of water upon the earth to destroy all flesh. Noah in obedience did as God commanded and the Lord honored him. With him God established His covenant (Gen. 6:18). How did Noah respond to God's call? Very favorably. He built the Ark as God commanded; took on the ark his wife, his three sons and their wives, and two of every sort of living thing of all flesh to keep them alive. Each set of living things that he brought on board consisted of a male and a female. Because of Noah's obedience, the Lord blessed Noah and his sons after the Flood was over. He said unto them, "Be fruitful and multiply and replenish the earth, and the fear of you and the dread of you shall be upon every beast of the earth, every fowl of the air and upon everything that moveth upon the earth; and upon all the fishes of the sea" (Gen. 9:1–2). Noah lived 950 years. He was 600 years old before the Flood and lived 350 years after the Flood. God also instituted the dispensation of human government through Noah in the new world (Gen. 9:1–6).

A View of Abraham

Based on Gen. 12:1–3 the Lord said unto Abram (Abraham), "Get thee out of thy country and from thy kindred and from thy father's house unto a land that I will show thee. And I will make of thee a great nation, and I will bless thee, and make thy name great; and thou shalt be a blessing. I will bless them that bless thee, and curse him that curseth thee, and in thee shall all families of the earth be blessed." This was God's call. Now let's see how Abraham responded. He responded positively. Abraham departed as the Lord had spoken unto him. And Lot, his nephew, went with him. Abraham was seventy-five years old when he departed Haran. Abraham took Sarai, his wife, and Lot, his brother's son and all their substance that they had gathered and the soil that they had gotten in Haran. And they went to the Land of Canaan (Gen. 12:1–4). God didn't have any difficulties with Abraham. When He called him, Abraham just said "yes." Great was Abraham's faith and faithfulness. And the Lord justified him through his faith.

A View of Isaiah, the Volunteer

The following is an account of Isaiah's call based on Isa. 6:8–11. "Also I heard the voice of the Lord saying whom shall I send, and who will go for us? Then said I here am I, send me. And He said, go and tell this people, hear ye indeed, but understand not; and see you indeed but perceive not. Make the hearts of this people fat and make their ears heavy and shut their eyes; lest they see with their eyes, and hear with their ears, and understand with their heart, and convert and be healed. Then said I, Lord how long? And he then answered until the cities be wasted without inhabitant, and the houses, without man, and the land be utterly desolate." From the Book of Isaiah, we learn that Isaiah heard the voice of the Lord saying, "Whom shall I send, and who will go for us?" He volunteers before the sovereign Lord and says, "Here I am; send me." What courage. Isaiah's response gave tremendous glory unto the Most High God. All of us should be like Isaiah and say, "We're available, Lord. Make me a vessel through which you work."

A View of Moses

The Lord said unto Moses, "I have surely seen the affliction of my people which are in Egypt, and have heard their cry by reason of their taskmasters; for I know their sorrows. And I am come down to deliver them out of the hand of the Egyptians and to bring them up out of that land unto a good land and a large, unto a land flowing with milk and honey. Come now therefore, and I will send thee unto Pharaoh, that thou mayest bring forth my people the children of Israel out of Egypt." God appeared unto Moses in the flame of fire out of the midst of a bush and made his call.

Let's take a look at Moses' response. Moses said unto the Lord, "Whom am I that I should go unto Pharaoh, and that I should bring forth the children of Israel out of Egypt? And when I go before the children of Israel and they ask me the name of the God of their fathers, what shall I say unto them?" God answered and said, "Thus shalt you say unto the children of Israel, *I am* sent me unto you." "I am" signifies that God would be Moses' source of strength and power. He will be a supplier of all needs. The Lord said unto Moses in Gen. 3:12, "Certainly I will be with thee and this shall be a token unto thee that I have sent thee." But Moses said, "O my Lord, I am not eloquent but I am of a slow tongue." But the Lord said to Moses, "Who hath made man's mouth? Who maketh the dumb, or deaf, or the seeing, or the blind? Have I not made them? Now go and I will be with thy mouth and teach thee what to say." Yet Moses had excuses. He even asked the Lord to send someone else. (Gen. 3 and 4). However, the Lord prevailed. When the Lord calls, just say "yes." In doing so you will avoid complications.

A View of Gideon

Let's tune into the dialogue between God and Gideon based on Judges 6:14–18: And the Lord looked upon him, and said, "Go in this thy might, and thou shalt save Israel from the hand of the Midianites: Have I not sent thee?" And he said unto the Lord, "Wherewith shall I save Israel? Behold my family is poor in Manasseh and I am the least in my father's house." "Surely I will be with thee and thou shalt smite the Midianites as one man." Gideon had low self-esteem. But God wasn't searching for ability; instead he was looking for Gideon's availability. God uses ordinary people to do extraordinary jobs. Glory be to God because it is He who performs the miraculous through vessels of honor. Therefore, when the Lord calls you, just say "yes." Think not about who you are or what you are; what you have or what you don't have. But concentrate on the confirmed call of God and He will direct your path. You will have a path of peace if you cast your cares on him because the Lord cares.

A View of Jeremiah

God even uses children. Let's listen to His sovereign call of Jeremiah based on Jer. 1:4–9. "Then the word of the Lord came unto me saying, Before I formed thee in the belly I knew thee, and before thou camest forth out of the womb I sanctified thee, a prophet unto the nations. Then said I, Ah, Lord God! Behold I cannot speak: For I am a child. But the Lord said unto me, Say not I am a child: For thou shalt go to all that I shall send thee, and whatsoever I command thee thou shalt speak. Be not afraid of their faces, for I am with thee to deliver thee, saith the Lord. Then the Lord put forth His hand and touched my mouth. And the Lord said unto me, Behold I have put my words in thy mouth." From this chapter we can clearly see that the Lord uses children as well as adults as vessels. However, in each call God says, "I will be with thee." And He fulfills His promises.

A View of Nehemiah

God put in Nehemiah's heart a burning desire to rebuild the walls of Jerusalem. He had such a strong sense of divine destiny for the nation that he gave up a favored place in the court of Persia in order to suffer with the returning exiles in the rebuilding of the Holy City. Nehemiah was patriotic, prayerful, perceptive, persistent, and persuasive. God gave him favor with the king. Though there was much opposition to the reconstruction of the walls, Nehemiah and his people prevailed because he knew God was with them (Neh. 4:13–23). Nehemiah had a very difficult assignment. But he feared not because he knew God would fight the battle for him and his people. His response to God's reconstruction call was very admirable. His soul said, "Yes, Lord." Among earthly men Nehemiah is an example to follow because he demonstrated godly character. He believed God. He obeyed God and he denied himself earthly joys to follow the Lord. He left his comfort zones in Persia to risk his life in pursuit of the joy of doing that which was pleasing unto the Lord. What a godly servant!

A View of Jesus, the Messiah

Because of Adam's and Eve's sin, the world needed a sinless Messiah who could reconcile us back to God. Because of man's fall from grace, he acquired a sin nature, which separates him from the Lord. It is written in John 3:16: "For God so loved the world that He gave His only begotten Son that whosoever believe in Him should not perish but have everlasting life." So Christ's mission was to do God's will (John 6:38); to save sinners from eternal damnation (Luke 19:10); to bring in everlasting righteousness (Dan. 9:24); and to destroy Satan's work (1 John 3:8). Jesus has not yet completed all of His work. However, he has begun it. Giving His life as a ransom for man was his basic task. Jesus ***gave*** his life because of His love and mercy on us. "He was led as a sheep to the slaughter and like a lamb dumb before the shearer so opened He ***not*** His mouth. In His humiliation His judgement was taken away" (Acts 8:32–33). Men were redeemed and now we have hope in Jesus for eternal life. We thank God for the craftsmanship He has performed in other vessels, but above all we thank Him for Jesus, who brought joy and hope to a lost world. Let us then keep our eyes focused on the Lord Jesus, the bright and morning star. And when the Lord calls you, just say "yes."

Macrowisdom and Microwisdom Leadership

"If any of you lack wisdom, let him ask of God, that giveth to all men liberally, and upbraideth not." James 1:5

XIX

The Macrowisdom and Microwisdom in Leadership

"Say unto wisdom thou art my sister; and call understanding thy kinswoman" (Prov. 7:4). According to this Scripture, understanding is a component part of wisdom, and they are both inseparable tables. Therefore, one cannot dine at the Table of Wisdom without the inclusion of the Table of Understanding. God says in Prov. 4:7, "Wisdom is the principal thing; therefore get wisdom; and with all thy getting get understanding." Now in this chapter we're going to focus our eyes on macrowisdom and microwisdom in leadership. But before we begin our illuminating discussion, let us define "wisdom," which is the key word. Ready?

The *Random House Dictionary of the English Language* tells us the following about wisdom: "Knowledge of what is true or right coupled with ***just*** judgment as to action; discernment or insight." In short wisdom is knowledge guided by understanding. From whence does it come? It comes from the Most High, Omniscient God who has invited men to sit and dine at His Table of Wisdom. God says in James 1:5, "If any of you lack wisdom, let him ask of God, that giveth to all men liberally, and upbraideth not; and it shall be given him." Furthermore, the Lord tells us in Isa. 33:6, "Wisdom and knowledge shall be the stability of thy times, and strength of salvation." In other words, wisdom and knowledge will yield stability, but lack of it will yield instability.

My friends, "The Lord by wisdom hath founded the earth; by understanding hath He established the heavens" (Prov. 3:19).

Now, I know you are waiting to hear about macrowisdom and microwisdom in leadership. So I am going to take you out of the suspense zones at this moment. In the subsequent paragraphs, we shall talk about macrowisdom, microwisdom, and I shall give you some biblical illustrations of each. How about that? Are you ready for the cruise now?

Macrowisdom

It's the application of knowledge in a situation that is large enough to be studied and identified with the naked eyes. One does not have to take a microscopic look at the subject. Instead the astronomer uses macrowisdom and common sense. Don't forget "common sense is not a common thing." An example of the application of macrowisdom: Your school has four hundred students and only nineteen classrooms. In each class there are twenty students. Twenty of the four hundred enrolled students are part-timers. That means there are three hundred and eighty full-time students and twenty part-time students. Classes begin at 8:30 A.M. and end at 4:00 P.M. There is a lunch break from 11:30 A.M.–1:30 P.M. Last class is from 2:00 P.M.–4:00P.M. The morning classes last one hour but the afternoon classes last two hours.

In order to graduate at the end of the school year, those twenty part-time students need to take two of the morning classes, which only last one hour each. Since there is no more space in the classrooms during the regular class periods, what shall we do? Hint: The cafeteria director will also allow the students to come and dine from 1:40 P.M.–2:10 P.M. if they desire. What do we do? We use macrowisdom and common sense. We schedule the part-time students' classes from 11:30 A.M. to 12:30 P.M. Then we give them a five-minute break period. Afterwards we schedule their second class from 12:35 P.M.–1:35 P.M. This is unquestionably usage of macrowisdom, since the situation is large enough to be seen with the naked eyes. All one has to do is use the given knowledge guided by understanding.

Microwisdom

It's the application of knowledge in situations that are too small to be seen clearly and understood by the naked eyes. The soul cries to the hills for wisdom and understanding. In His word the Most High God said, "If any man lacks wisdom, let him ask the Lord who giveth wisdom liberally and upbraideth not." The following is an example of the application of microwisdom. Mr. and Mrs. Doe are born-again believers, who have been servants of the Most High God for fifteen years. They have four children to support, mortgage payments, and other primary necessities of life. Because of Mr. Doe's previous illness, the family acquired an enormous medical bill.

Recently, both Mr. and Mrs. Doe lost their jobs, which provided for their family's needs. Now they're unable to pay their mortgage and other bills. They're constantly harassed by bill collectors. They seem to have trouble at every angle of their lives. Because of the unemployment problem of the city where they live, jobs are very difficult to find. One cannot easily find a job but can easily find himself without one. Last week they took their 10 year old to the hospital. Today they learned that he has been diagnosed with leukemia. With these facts in mind, the Doe family knows that the only hope they have is in Christ Jesus.

One needs to apply microwisdom in situations like these. The Old Testament teaches, "Many are the afflictions of the righteous, but the Lord delivers them out of them all." The Lord Jesus is the bridge over troubled water. He stills the storms and calms the angry sea; He can still your storms of life; and He will calm the angry seas of trouble in your life if you confide in Him. Jesus is the Prince of Peace and He merits our trust. Whoever experiences a trial similar to that of the mock trial of the Doe Family must remember that microwisdom needs to be applied. The Doe Family prevailed because they remembered the majestic words of King David, a spiritual astronomer: "I have been young, and now am old; yet have I not seen the righteous forsaken, nor his seed begging bread" (Ps. 37:25). "Beloveth, according to the Word of God, the Lord hath his Way in the whirlwind and in the storm, and the clouds are the dust of His feet" (Nahum 1:3).

Biblical Illustrations of the Usage of Macrowisdom

The Three Wise Men from the East

The Scripture tells us that Jesus was born in Bethlehem of Judea in the days of Herod. Behold three wise men from the East came to Jerusalem searching for the child. Having seen the star in the East, they came to worship Him. Herod, being troubled about the news of the birth of the Messiah, gathered all the chief priests and scribes of the people together to demand of them where the child should be born. They told Herod that the child would be born in Bethlehem in the land of Judah. Then Herod called the three wise men to a private meeting to inquire secretly of them what time the star appeared. Afterwards he sent them to Bethlehem and said, "Go and search diligently for the young child, and when ye have found him, bring me word again that I might come and worship him also."

When they had heard the King, they departed and the star that they had seen in the East went before them until it came and stood over the place where the young child was. The wise men rejoiced. They went into the house and saw the young child with Mary, his mother. Then they fell down and worshiped him. However, using macrowisdom and also being warned of God in a dream that they should not return to Herod, they returned to their own countries another way. They were wise men indeed. *Macrowisemen.* Had they returned to Herod, he would have gone to Bethlehem to slaughter the child, not to worship him. When he saw that he had been mocked by the *macrowisemen,* he became very angry and sent forth and slew all of the children that were in Bethlehem and in all of the coast thereof from two years old and under (Matt. 2:1–12).

The Parable of the Ten Virgins

Here's another demonstration of usage of macrowisdom. Ten virgins took their lamps and went to meet the bridegroom. Five of them were wise and five were foolish. The foolish ones took their lamps but took no oil with them. But the wise virgins took oil in their vessels with the lamps. At midnight a cry was made, "Behold the bridegroom cometh; go ye out to meet him." Then all those virgins arose and trimmed their lamps. And the foolish said to the wise, "Give us some of your oil. Our lamps are gone out." But the wise answered saying, "Not so lest there be not enough for us and you; but go ye rather to them that sell and buy for yourselves." While they went to buy, the bridegroom came and they that were ready went in with him to the marriage and the door was shut. (Matt. 25:1–13).

A Mother in Israel during the Time of a Crisis

Believing that Joab and his men want to destroy Israel and the inheritance of the Lord, she uses her God-given macrowisdom to confront him and conduct this dialogue: "I am one of them that are peaceable and faithful in Israel; thou seekest to destroy a city and a mother in Israel; why wilt thou swallow up the inheritance of the Lord?" And Joab answered and said, "Far be it, far be it from me, that I should swallow up or destroy. The matter is not so but a man of Mount Ephraim, Sheba the Son of Bichri by name, hath lifted up his hand against the King, even against David; deliver him only and I will depart from the city." And the woman said unto Joab, "Behold his head shall be thrown to thee over the wall." Then the woman went to all the people in her wisdom and they cut off the head of Sheba the son of Bichri, and cast it out to Joab. And blew a trumpet and they retired from the city, every man to his tent. And Joab returned to Jerusalem unto the King (2 Sam. 20:14–22). This mother in Israel applies macrowisdom in order to free her city from those who sought after Sheba, the son of Bichri. Sheba was a destructive man.

Though I have only given you a few samples, there are many biblical illustrations of the usage of macrowisdom. The Old and New Testaments contain an enormous amount. Now let us go to the house of microwisdom and see what the Holy Spirit has in store.

Biblical Illustrations of Usage of Microwisdom

Based on James 1:5 the Most High God is indisputably the source of all wisdom. He applies it in His works and also gives a liberal portion to men and angels. He gives us macrowisdom and he gives us microwisdom. Since microwisdom is the highest form of wisdom given to man, I chose to include the name of the Most High God as an example. Prov. 3:19 declares that, "The Lord by wisdom hath founded the earth; by understanding hath He established the heavens." This is a demonstration of God's highest form of wisdom. By wisdom, God created man; by wisdom God created angels and all things. Finally, through wisdom of the Most High God came the incarnation of our Lord and Savior Jesus Christ. The Virgin Mary was God's earthly vessel.

King Solomon and the Two Women's Dispute over a Child

During King Solomon's reign, he had the experience of using microwisdom to judge two women's dispute over which of the two was the true mother of the living child. The two women lived together and they both had a son. However, one of the women informed the king that one of the sons had died because the mother lay on him during the night. "Then she rose at midnight and took my son from beside me while I slept and lay it in her bosom, and brought her dead son and lay him beside me." After listening to both of them, King Solomon said, "One of you say this is my son that liveth and thy son is the dead; and the other say nay, thy son is the dead and my son is the living. Bring me a sword," and they brought a sword before the King. And the king said, "Divide the living child in two; and give half to the one and half to the other." Then spake one woman, "O my Lord, give her the living child, but in no wise slay it." But the other said, "Let it not be neither mine or thine, but divide it." Then, the king answered and said, "Give her the living child, and in no wise slay it; she is the mother thereof." That's microwisdom! Bravo! Bravo! Praise the Lord! (1 Kings 3:16–28).

The Prophet Daniel

Based on Dan. 2:1–45, Daniel used microwisdom to interpret King Nebuchadnezzar's dream. It required intensive prayer. God revealed the dream and gave Daniel the interpretation. God used Daniel's gift to bring honor and glory to His name. Later Daniel uses his microwisdom to interpret the supernatural handwriting on the wall of King Belshazzar's court.

Elijah on Mount Carmel

It was really a contest between the living God of Heaven and earth and between Baal, the idol god. Elijah said to the people, "If the Lord be God, follow him; but if Baal, then follow him." The people remained silent. Then Elijah said, "Baal's prophets are four hundred and fifty men; let them therefore give us two bullocks and let them choose one bullock for themselves, and cut it in pieces, and lay it on wood and put no fire under: and I will dress the other bullock and lay it on wood, and put no fire under. And call ye on the name of your gods and I will call on the name of the Lord; and the God that answereth by fire let him be God." And all the people answered and said, "It is well spoken." Elijah's God answered by fire and the people saw it and they fell on their faces and they said, "The Lord He is the God." Then Elijah said, "Take all the prophets of Baal and don't let any of them escape." They took them and Elijah brought them down to the brook Kishon and slew them there (1 Kings 18:15–46).

The Prophet Job

I am sure most of you readers are familiar with the trials, suffering, patience, and faith of Job. But for those who are not, please read the Book of Job in the Old Testament of the Holy Bible. Briefly, God had blessed Job with faith, a wife and family servants, cattle and much prosperity. However, Satan, the adversary of God and man, was allowed to test him. He brought sickness upon him, body sores and the like; Job lost cattle, lost his children, and he said, in wisdom, "The Lord giveth and he taketh away." He also said, "I brought nothing into this world and nothing shall I take out." Job's trials and calamities were so hard that his wife suggested that he curse God and die. He told her, "You talk like a foolish woman." Though his trials were very hard, God prevailed in the life of Job.

Let's tune in now to Job's microwisdom and statements of faith found in Job 19:25–28: "For I know that my Redeemer liveth, and that he shall stand at the latter day upon the earth. And though after my skin worms destroy this body, yet in my flesh shall I see God: whom I shall see for myself, and mine eyes shall behold, and not another; though my reins be consumed within me." But ye should say, "Why persecute we him, seeing the root of the matter is found in me?" Job's wisdom obviously came from God in his mercies as he looked to the hills for help. No one can go through such trials without experiencing defeat unless God fights the battle for him. To God be the glory and honor in the life of Job. Amen. As leaders we must learn to trust in the Lord and lean not on our own understanding.

Test and Measurement

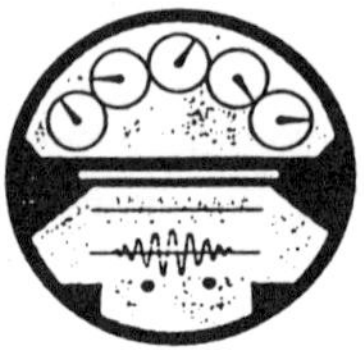

"Your faith being much more precious than gold
that perisheth, though it be tried with fire,
might be found unto praise and honor and
glory at the appearing of Jesus Christ."
1 Peter 1:7

XX

Tests and Measurements

God is the divine tester and His people are His testees. He tests and measures them. After their tests, they return the test papers. Some score 100 percent and some fail miserably. Some fail miserably even when God's tests are very, very easy! On God's tests there is no middle ground. You may choose to pass or fail the exam. Now let's explore the meaning of test and measurement.

Test

It is a means by which the quality or genuineness of anything is determined. In a spiritual sense it is a trial that manifests one's real character.

Measurement

It is the extent or size ascertained by measuring. We may also conclude that it is a measured dimension, which tells us the spiritual width, length, depth, and height. If Christ dwells in one's heart by faith, He is rooted and grounded in love. Therefore, He will be able to comprehend with all saints "what is the breadth and length and depth and height." And He will "know the love of Christ, which passeth knowledge" (Eph. 3:18–19). Perhaps you might ask the following three questions:

1. Since God knows all things, why does He test man?
2. What are the purposes of the tests?
3. What sort of tests does God administer or allow us to experience?

I shall answer these questions for you in the subsequent paragraphs. Let us begin.

Question 1 Since God knows all things, why does He test man?

Answer God doesn't test us to learn about us because He knew us before He even created the foundation of the world (Eph. 1:4). Instead He tests to reveal truths and secrets to the testees—truths about their motives for serving Him, truths about their character, their faithfulness, their sincerity, their weaknesses, their patience, etc. A schoolteacher tests his students to *evaluate* their knowledge of the lessons he has taught them. But God tests His people to *teach* them lessons and that they might see the hand of God at work as He strategically and successfully brings them through trials of life. If it were not for the divine intervention of God's Holy Spirit, we would flunk all of the Lord's exams. The Lord says in 1 Pet. 1:7, "That the trial of your faith being much more precious than of gold that perisheth, though it be tried with fire, might be found unto praise and honor and glory at the appearing of Jesus Christ?" If we trust God, we will surely score 100 percent on our tests of life because He never fails. God even makes his strength perfect in our weaknesses. Unlike men, God allows us to call on Him during the time of testing. In Jer. 33:3 God says, "Call on me and I will show thee great and mighty things."

Question 2 What are the purposes of the Lord's tests?

Answer In short, the primary purposes of the tests of the Lord are: A. To test our obedience (Gen. 3:1–8); B. To confirm our faith in God (Gen. 22:1–8); C. To learn the will of God (Jud. 6:36–40); D. To refute Satan's claims (Job 1:6–22).

God doesn't give midterm or semester exams. Instead He

gives us daily tests. Some are more difficult than others, but through His grace, we are more than conquerors. Thanks be to the Most High God for His redeeming love and grace. Whenever God allows one to be tried or tested, He always gives him sufficient grace to weather the storm. And He says to him, "Peace be still and know that I am God." Let's be still and listen to God because one day we're all going to take our final exam.

Question 3 What sorts of tests does God administer or allow us to experience?

Answer Physical, supernatural, and spiritual are excellent choices of adjectives, which vividly describe the nature of God's tests. Let me quickly give you some examples from the Word of God.

Example A: Physical Test

A test is physical when we need to utilize physical strength, intelligent strategy, and a sound mind in order to obtain victory. Though the test is physical, relying on your muscles alone will never win the battle. The testee must always remember that the battle is the Lord's and He will give him the victory. Now how about a telescopic look at little David and Goliath, the Philistine giant. David was just a youth and his physical strength obviously could not compare with that of Goliath, the giant who was Israel's enemy and an enemy to the people of God. Many feared the giant, but Little David didn't. David prevailed over the Philistines with a sling and with a stone, knowing that the battle was the Lord's (1 Sam. 17:39–51).

A test may also be considered physical when the body is ill and needs healing. The testee relies on God for healing whether He sends it through medicine, through supernatural divine healing, or gradually through natural means. Many lessons are learned on a sick bed. The greatest one is the lesson of cause and effect. Sometimes the victim discovers that he is the root cause of his illness. All he needs to do is abstain from drinking, smoking, improper dieting, or deviant lifestyles. God gives his sons, and daughters, the power to become the men and women He desires them to become. Remember sickness is the effect, but the cause always precedes the effect. Rule of thumb: There can be no effect without a preceding cause. That's law; that's logic. And that's God's order. And that settles it. Amen!

Example B: Supernatural Test

Whenever there is direct influence or action of God on earthly affairs, it is supernatural. Supernatural is being beyond that which is natural. In order to deliver His enslaved people out of Egypt land, God used Moses as a vessel and gave him supernatural power. Moses believed God and passed his exams. For example, Moses stretched out his God-given rod and the Red Sea was parted and the Hebrew children walked through, but Pharaoh's army was drowned. Many supernatural acts of God took place in Egypt before the Exodus of the Hebrew slaves (Exod. 7–11). Beloved, be not deceived, God does not perform miracles to entertain us. He does them at the proper time and for His glory and honor. God still uses supernatural means today to lead us out of the paths of danger, to heal our sickness, etc., because our God is the same yesterday, today and forever. In other words, He is an unchanging God.

Example C: Spiritual Test

Let us first define spiritual. What is it? Spiritual is that which pertains to one's spirit as the seat of his religious or moral character. That which is spiritual is holy and immaterial. Therefore, a spiritual test is one that is used to test a man's spirit and faith in God. The Most High God allows us to experience these tests that we might draw nigh to Him. He is the only one who can deliver us. Let's take a look at the prophet Daniel in the lion's den of Babylon. In 605 B.C. Nebuchadnezzar, the wicked king of Babylon, invaded Palestine and brought King Jehoiakim to his knees and carried off hostages, among them Daniel and his three associates (Hananiah, Mishael, and Azariah). Their names were changed while in captivity. Be that as it may, Daniel refused to honor any idol god. He would only worship the Most High God, the Holy One of Israel. King Darius and other wicked leaders of Babylon tried to break Daniel's spirit by putting him in the den of lions. But Daniel's spirit was not broken; instead he became stronger. God sent an angel to close the lions' mouths. Daniel was delivered and King Darius wrote unto all people, nations, and languages that dwelt in all the earth. He made a decree that in every dominion of his kingdom men tremble and fear before the God of Daniel (Dan. 6:1–28).

Before I conclude this chapter, I am going to point out some areas where these tests are manifested. They are prevalent in marriages, in friendships, in courtships, in parenting, in employment or careers. God also uses these tests to test the spirit of an individual and to test the judicial system of nations.

Manifestation of a Marriage Test

In many marriages couples make a vow to stick together in poverty and in wealth, in sickness and in health, and to depart from each other only at death. When things go well, many keep their promises. But when the road gets rough and things don't go as expected, insincere couples' marriages fall apart. Let's take a look at the prophet Job and his wife. Her true character was revealed to this once very prosperous man on his bed of affliction at the time he needed her support the most. That ungodly woman told Job to curse God and die. What a wife! What a wife! And what sympathy! (Job 2:9–10).

Manifestation of a Friendship Test

A true friend is an intimate associate on whom you may rely. He is one who sticketh closer than a brother. When a so-called friend who is able to assist you, fails you in time of need, you have witnessed the manifestation of a test of friendship. The Word of God says, "A friend loveth at all times and a brother is born for adversity" (Prov. 17:17). Try Jesus, my friend. He is the only friend that sticketh closer than a brother (Prov. 18:24). His love is unfailing and He is unchanging.

Manifestation of a Parental Test

Many parents labor very hard in order to provide their children with necessities and comforts of this life. Yet some of the children flunk the simplest parental tests: "Go to the store," "go to the post office," or "take out the garbage," even "drop me off at the doctor's office for my appointment." Many children flunk these parental exams even when they have nothing else to do. The only passwords the children know how to say to their parents are "Give me this or give me that." They show no love for such loving parents. The only love they show is a "love to manipulate." However, it is the parents' duty to train and discipline their children that they might perform better on parental exams. Proverbs 22:6 supports this.

Manifestation in Courtship Test

Many tests take place during the courtship of a couple that they might discern if they are right for each other. They also learn about their faithfulness, truthfulness. It's decision-making time. The Bible says, "Do not be unequally yoked together with unbelievers" (2 Cor. 6:14).

Manifestation of Spiritual Tests on Your Job Site

Eph. 6:12 tells us, "For we wrestle not against flesh and blood but against principalities, against powers, against the rulers of the darkness of this world and against spiritual wickedness in high places." Sometimes Satan and his powers of darkness manifest themselves in your jobs through human vessels. He's only there to cause confusion, to spread lies, and to make you uncomfortable. However, you must remember that you're going to weather the storm because God sent you there. You have security in Jesus. Just do as the Holy Spirit of God directs you and that's victory. If God be for us who can be against us (Rom. 8:31).

Manifestation of God's Test in the Judicial Justice System of Nations

Thinking democratically and godly, I'm going to say here that the judicial system is all about the just laws of the land. When these laws are carried out, we see the beauty of a democracy. "No one is above the law" sounds beautiful on paper, but is it true? Unfortunately, there is the discretionary factor in law enforcement throughout the world. Whenever the discretionary factor is applied, justice hides its beautiful face and then we see the display of the ugly face of injustice. However, God sees all that goes on in the nations of the world. He has accurate records above. This is why he says, "Pray for kings and those in authority. It is not the will of God that any should perish" (1 Tim. 2:1–2).

How wide is the forest? How deep is that ocean? How tall is that tree? How long is the journey to heaven? The answers, my friend, are written in the winds. Only God knows the answers. He tests and He measures. The forest is as wide as He says it is; the ocean is as deep as He says it is; the tree is as tall as He says it is; and the journey to heaven is as long as He says it is. One thing is sure, as we walk humbly with our God up life's highway, He will give His children revelation of width, length, depth, and height. We cannot go through life without the knowledge of this spiritual math. God knows the answers. He tests and measures. Ask Him. He invites you. In His word He says "seek and ye shall find. Knock and the door shall be opened" (Matt. 7:7). Always remember if the jurisprudence of a nation fail to honor God, then the nation will surely collapse. "Righteousness exalteth a nation but sin is a reproach to any people" (Prov. 14:24).

Empires come and go! Empires come and go! History records it! History records it! When you read about the fall of ancient empires, you'll notice that they all had one thing in common at the time of their fall: *Ungodliness.* In other words, *mass wickedness!* God is a holy and just God, and He will surely punish the sins of nations when they don't repent. The Holy Bible teaches us that "the wrath of God falls upon the children of disobedience."

How wide is the forest? How deep is that ocean? How tall is that tree? How long is the journey to heaven? The answers, my friends, are written in the winds . . . written in the winds! Only God knows the answers. He tests and measures.

Jacob´s Ladder or the Bottomless Pit?

Church leader, the choice is yours. Which ladder will you and your flock climb? Will you ascend Jacob´s ladder into heaven or descend Satan´s ladder into hell? The Lord Jesus, the captain, is watching you.

STARTING POINT: The Earth

Satan´s Ladder
Psalm 139:8

XXI

Up Jacob's Ladder or Down the Bottomless Pit's Ladder?

Church leader, the choice is yours. Which ladder will you and your flock climb? Will you ascend Jacob's ladder into heaven or descend Satan's ladder into hell? The Lord Jesus, the captain, is watching you. And the choice is yours. So what are you going to do, shepherd? You're on the planet Earth, God has given you sheep to feed, and there are two great ladders standing before you. One leads to eternal life in a glorious place called heaven where the streets are even paved with gold. The other leads to eternal destruction in an evil place called hell where the wicked get no rest during the day nor at night.

Based on Gen. 28:12–13, Jacob saw that heaven-bound ladder. He said he saw a ladder set up on the earth, which had a top that reached heaven. Moreover, Jacob saw angels of God ascending and descending on it. And behold, the Lord stood above it, and said, "I am the Lord God of Abraham thy father, and the God of Isaac." But King David, a spiritual astronomer, had a different vision. David said unto the Lord, "If I ascend up into heaven, thou art there; if I make my bed in hell, behold thou art there" (Ps. 139:8).

Behold, I see two ladders in this Scripture: One ladder ascends into heaven, and the other descends into hell. If you lead the sheep in the path of righteousness, both you and they are climbing Jacob's ladder. If you show your sheep that Jesus is the way, the truth, and the life, both you and they are climbing up Jacob's ladder (John 14:6). But if you lead yourself and your sheep astray, you're descending the bottomless pit ladder, which leads to hell. Keep your eyes fixed on Jesus, our spiritual astronomy, and He will guide you and empower you through the Holy Ghost.

Now I know you have some questions. So I'll take the first one now.

1. "Can a wicked shepherd lead God's sheep astray?" Thank you. That's an excellent question. The answer is a resounding No! No! No! The Lord Jesus says in the word, "My sheep know my voice and they follow, but a stranger they will not follow, but will flee from him: for they know not the voice of strangers" (John 10:4–5). God watches over the redeemed and He says "none shall be plucked out of His hands." Now I know you have a lot of questions, but I am only going to answer three more because the Holy Spirit is leading me to conclude this book and write a brief summary.

Next question: "How can the sheep know whether or not their leader is sent by God?" The New Testament teaches us that by their fruit ye shall know them. Now the fruits of the spirit are love, joy, peace, gentleness, meekness, goodness, patience, temperance, faith; against such there is no law. You also need to remember that wolves can also come in sheep's clothing, but inwardly they are ravening wolves. Therefore, we need to ask our all-knowing God to give us a discerning of spirits. (This answer is based on Matt. 7:15–17 and Gal. 5:22.)

Next question: "How does one climb that ladder which leads to eternal life?" The answer, my friend, is found in Matt. 7:13–14: "Enter ye in at the straight gate; for wide is the gate, and broad is the way, that leadeth to destruction and many there be which go in thereat. Because straight is the gate and narrow is the way, which leadeth unto life, and few there be that find it."

Remember you're a spiritual astronomer. As you climb up Jacob's ladder, just keep your eyes fixed on Jesus, the bright and morning star. If you obey God, you will surely enter the promised land. Now I'll take only one more question. Your question please.

"When a church leader discovers that there are wandering sheep and Judas sheep in his congregation, what should he do?" That's an outstanding question, and I'm going to give you a simple outstanding answer. It's all about stewardship. The Lord tells us in the New Testament to be good stewards over that which He gives us. God has made you an under-shepherd in His house. Therefore, you must protect the sheep from the lions who come there to destroy *them* and also the constructive work going on in God's house. If the wandering sheep and the Judas sheep choose not to repent before our living God, and if they choose not to listen to you, then your only alternative would be to excommunicate them (Matt. 8:15–17). You must protect your ministry. Don't allow Satan, the adversary, any room. If you allow him an inch, he will take a mile.

In concluding this chapter, let me reiterate: church leader, the choice is clearly yours. Which ladder will you climb with your flock? Will it be Jacob's ladder, which ascends into heaven, or will it be the ladder that descends into the bottomless pit of hell? The prophet Nehemiah and his people heroically rebuilt the walls of physical Jerusalem. But we, the modern church leaders, have the unique challenge of constructing a retaining wall around spiritual Jerusalem. If we fail to construct the wall, then the adversary of God and man will prevail in spiritual Jerusalem. He will invade us and we will surely become annihilated. However, if we build the wall around spiritual Jerusalem, the Lord Jesus will surely prevail. Neither the mercenaries of Satan nor those of men will be able to enter spiritual Israel. When we obey the Most High God, the gates of hell cannot prevail against us. Hallelujah! Hallelujah! Hallelujah!

Summary

Spiritual astronomy is unquestionably the heartbeat of dynamic Christian leadership. In order to become a spirit-filled Christian leader astronomer, one must march to the exciting drumbeats of God's Holy Spirit. Christian leadership is not Christian leadership because we *say so,* but it is Christian leadership when we *act so.* "Faith if it hath not works, is dead" (James 2:17).

If we're going to utilize the Word, *Christ* in leadership, then we must focus our eyes on the Lord Jesus Christ, who is the main attraction. Jesus declares in Rev. 22:16, "I am the root and offspring of David and the bright and morning star." Beloved Saints of God, let us follow the star and He will surely lead us to the promised land.